SPACE AND TIME

Richard Swinburne

Second Edition

First Edition 1968
Second Edition 1981

Published by
THE MACMILLAN PRESS LTD
London and Basingstoke
Companies and representatives
throughout the world

Printed in Hong Kong

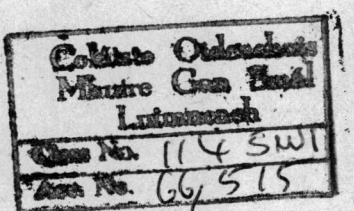

British Library Cataloguing in Publication Data

Swinburne, Richard
Space and time – 2nd ed.
1. Space and time
I. Title
114   BN632

ISBN 0–333–29072–0

SPACE AND TIME

Contents

Preface to the Second Edition

The first edition of *Space and Time* was published in 1968. It was intended both as a guide for students to the scientific and philosophical literature of the subject and as a vehicle of my own philosophical views. In the twelve years which have now passed since I finished writing the first edition, there have been very considerable advances in astronomy, physical cosmology, and particle physics, as well as a great number of philosophical books and articles on space and time. As I find myself still in agreement with most of the detailed arguments of the first edition, I thought it worthwhile to produce a new edition, revised to take account of these scientific and philosophical developments.

On one crucial philosophical issue, however, I have come to hold a very different view from the unargued view implicit in the first edition. In 1968 I was a verificationist; I assumed that if no evidence of observation could in any way count for or against some statement, then necessarily that statement was empty of factual meaning. I now think that this is not so; a statement may make a meaningful factual claim even if evidence of observation can give no grounds for believing it to be true or for believing it to be false. In the new Introduction I give brief reasons for my change of view. It has necessitated a redescription of a few of my results at various places in the book.

I am most grateful to Mrs Yvonne Quirke for her patient typing of sections of the second edition.

Preface to the First Edition

I am most grateful to all who read or heard earlier versions of various chapters and produced helpful criticisms of them, and especially to Dr Elizabeth Harte, Dr Mary Hesse, Mr J. L. Mackie, Mr Roger Montague, Professor Alan White, and the late Mr Edward Whitley. They saved me from many silly mistakes and provided me with useful distinctions and arguments. I am also most grateful to my wife for reading the proofs.

Some of the material has previously appeared in articles in journals, although some of the conclusions which I reach in this book are substantially different from conclusions reached in the articles. I am grateful to the editors of the following journals for permission to use material from the articles cited: *Analysis* ('Times', 1965, 25, 185–91, 'Conditions for Bitemporality', 1965, 26, 47–50, 'Knowledge of Past and Future', 1966, 26, 166–72); *Philosophical Quarterly* ('Affecting the Past', 1966, 16, 341–47); *Philosophy of Science* ('Cosmological Horizons', 1966, 33, 210–14); and *Proceedings of The Aristotelian Society* ('The Beginning of the Universe', Supplementary Volume, 1966, 40, 124–38). In the references to these articles and to articles throughout the book, the first number indicates the year of publication of the journal, the second number the volume of the journal, and the third number gives the pages of the journal in which the article is to be found.

I have placed at the end of each chapter a bibliography of the most important works on the main topics of the chapter, from which the argument is developed. The bibliography also includes elementary works expounding the scientific theories or mathematical systems discussed in the chapter. Numbers in square brackets refer to the bibliography at the end of the chapter. Works referred to in the text which are important only for minor points in my argument are cited in footnotes but not included in the bibliography.

Introduction:
Necessity and Contingency

This book has two aims. The first and major aim is to analyse the meaning of various claims about Space and Time, to show which properties Space and Time must have as a matter of logical necessity, and to show what are the logically necessary limits to our knowledge of events in Space and Time. It thus considers such questions as whether Space must of logical necessity have three dimensions, what it means to say that an object is at rest, or whether it is logically possible that we should know as much about the future as we do about the past. The second and subsidiary aim is to describe, where a definite conclusion can be reached, the most general contingent properties of the Space and Time with which we are familiar, and the physical limits to our knowledge of events in Space and Time; or, when there is still scientific dispute, to set forward the rival scientific theories on the subject. It is thus for example concerned to describe theories about the age of Universe, and the limits to our knowledge of the Universe caused by the finite velocity of light.

The plan of the book is roughly as follows. In Chapters 1–7, I analyse the meaning of spatial terms and the status of propositions about Space, leaving unanalysed the meaning of temporal terms. In Chapters 8–11, I then conduct the analysis of temporal terms and the propositions about Time. In Chapters 12–15, I examine what are the physical limits to our knowledge of the Universe and what sort of conclusions we can reach about its general spatio-temporal character.

Considerations of what would be a proper length for the book unfortunately preclude discussing at any great length problems of space and time on the very small scale, such as whether distances and temporal intervals are divisible without limit. Such problems are much fewer and to my mind of much less philosophical interest than problems of space and time on the very large scale, and adequate discussion of the former would necessitate full treatment of Quantum Theory, just as adequate discussion of large-scale problems necessitates full treatment of the

Theory of Relativity. For the latter but not for the former I have found room.

For the benefit of any unfamiliar with the terms I must give a brief and necessarily very superficial exposition of the distinction between the logically impossible, the logically contingent and the logically necessary. To the initiates I apologise.

The distinction between the logically contingent and the logically necessary is an ancient one but it was first formalised explicitly by Kant [1] as his distinction between the synthetic and the analytic. Kant's formulation of this distinction suffers from various deficiencies and I will give a more modern one.

All statements are either logically contingent (synthetic or factual), logically necessary (analytic), or logically impossible. Any statement which is either logically necessary or logically contingent is logically possible. A logically impossible statement is one which entails a contradiction. For example, 'There are squares which do not have four sides' is logically impossible; for we mean by 'square' 'figure with four equal sides and for equal angles', and so the statement entails the claim that there are figures which both do and do not have four sides. An analytic or logically necessary statement is one whose negation is logically impossible. The negation of a statement is the statement which says that things are not as the first statement said. Thus the negation of 'There are men on the moon' is 'There are no men on the moon'. Hence 'All squares have four sides' is a logically necessary statement. So too is 'All bachelors are unmarried', 'Acceleration is the rate at which velocity increases', '$2 + 2 = 4$', and 'If there is a God, he is omnipotent'. We mean by 'God' a being who is omnipotent, omniscient, perfectly good etc. Hence 'it is not the case that if there is a God, he is omnipotent' is logically impossible, for it entails the claim that 'It is not the case that if there is an omnipotent being, there is an omnipotent being'. Logically impossible statements must be false, whatever the world is like; logically necessary statements must be true, whatever the world is like.

A logically contingent or factual or synthetic statement is one which does not entail a contradiction, and whose negation does not entail a contradiction. Among synthetic statements are 'Kangaroos live wild only in Australia', 'Julius Caesar crossed the Rubicon in 49 B.C.', and 'Silver dissolves in nitric acid'. There is no contradiction involved in any of these statements, nor in their negations. It may be false to claim that kangaroos live wild in parts of the world other than Australia, but I do not contradict myself in claiming this. Synthetic statements may be true or false; which they are depends on how the world is. Philosophers distinguish between sentences on the one hand and statements or propositions on the other. A sentence is a grammatically well-formed group of words. A statement is what is expressed by a sentence when the utterer of the sentence thereby asserts that something is the case (as opposed to, for example, issuing a command or asking a question). 'Socrates is a man' expresses the same statement as any sentence which has the same meaning as it. Thus it expresses the same statement as the Latin 'Socrates est homo' or the French 'Socrate est un homme'. Different sentences may thus express the same statement. The same words may, however, in the course of time acquire different meanings and so the same sentence may express different statements. 'Judge Smith is an indifferent judge' used to mean that Judge Smith is an impartial judge, but now it means that he is a judge who is not concerned about the issue of this case.

Given their current meanings, the words in a sentence determine which statement that sentence currently makes. If the words of a sentence with their current meanings are alone sufficient to ensure that that sentence makes a true statement, then that statement will be analytic. That the sentence 'all bachelors are unmarried' makes a true statement is guaranteed by the meanings of the words which it contains. If the words of a sentence with their current meanings are not sufficient to guarantee either the truth or falsity of the statement which it makes that statement will be synthetic. My inquiry is into the nature of statements about Space and Time as made by sentences uttered during the period of history in which I am writing by normal users of language or, in the case of sentences using technical scientific terms, by normal users of the scientific language.

We do not need to observe the world to find out whether or not a statement is analytically true. We need only reflect on the concepts denoted by the terms in the sentence making the statement; on the meaning of those terms. We can ascertain that the statements are analytically true independently of other experience of the world, that is *a priori*. Statements, the truth or falsity of which we can ascertain in this way, are termed *a priori* statements. It would seem also natural to suppose that we must use our experience of the world to find out whether or not synthetic statements are true. Only a study of animal ecology will tell us whether or not kangaroos live wild only in Australia; only a study of documents and monuments will tell us in which year Caesar crossed the Rubicon. Statements, the truth or falsity of which we can find out only by experience, are termed *a posteriori*. So it is natural to maintain that while all analytic statements are *a priori*, all synthetic statements are *a posteriori*. Several philosophers, and classically Kant, have, however, claimed that there are synthetic *a priori* statements — that is, statements informative about the world, the truth of which can be ascertained independently of experience of the world. Kant claimed that the statements of geometry and 'every event has a cause' were such statements. Most philosophers of the British empirical tradition have, however, denied that there are any synthetic *a priori* statements. Following the hints of Hume,[1] they have claimed that the distinction between the synthetic and the analytic coincides with that between the *a posteriori* and the *a priori*. There is not space to argue this issue, and so I will take for granted this coincidence, as seems initially reasonable. When discussing the views of Kant or others in his tradition on some statement which they claim to be a synthetic *a priori* truth, having stated their claim, I shall subsequently represent the claim as the claim that the statement was analytic or logically necessary.

A number of recent writers, and most famously Quine [2], have denied the applicability to our discourse of this distinction between the synthetic and the analytic. One reason which they have had for denying this is that in their view there is no clear criterion for when sentences do or do not have the same meaning and so for whether they make the same statement. A similar reason is that any precise definition of 'analytic' and 'synthetic' brings in notions like 'containing' or 'entailing' a 'contradiction', and that in their view there is no clear criterion for when one statement 'entails' another, or when we have a 'contradiction'. The Quinean attack however seems to me to fail because all these words either have an established use in ordinary language or can be defined by words which do have such a use. We often say of two sentences that they have the same meaning or that they say the same. If we could not distinguish whether two sentences did or did not have the same meaning, we could never translate from one language to another; and, despite the difficulties of translating some sentences into some languages, we often make perfect translations. True, whether or not two sentences mean the same may sometimes depend on the context in which these sentences are uttered, but that does not affect the main point. Likewise we can give clear examples of statements which 'entail' other statements or 'contradict' other statements, and these words can be defined by such expressions as 'denies what the first statement says' or 'would contradict yourself if you said this and denied that', which have established uses in

[1]David Hume, *Enquiry Concerning Human Understanding* (1st edn 1748) ed. L. A. Selby-Bigge; 2nd edn (Oxford, 1902) p. 25.

ordinary language. There is unfortunately no space in this book for a full discussion of Quine's attack on the distinction, and the book must assume that the Quinean attack can be dealt with adequately on the above lines. (For further discussion see [3] and [4].)

The examples which we have taken so far of analytic and logically impossible statements are ones where it is easy to see the logical status of the statements. In this book we shall be concerned with statements whose logical status is in no way obvious — e.g. 'no agent can bring about a past state of affairs' or 'there is only one Space'. How are we to show whether such a statement is analytic, synthetic or logically impossible? By inquiring whether either the statement which the sentence makes or its negation entails a contradiction. You can show that a statement entails a contradiction by deducing that contradiction from it. The only way to show that a statement does not entail a contradiction is to show that it follows from some other statement which does not entail a contradiction. Plenty of ordinary statements are to all appearances free of contradiction, and must be taken to be so in the absence of proof to the contrary. Thus 'Today is Friday', 'This wall is green', 'All swans are white', 'All students wear jeans' etc. are in this category. If you can show that a doubtful statement *q* follows from a statement or conjunction of statements *p*, which are to all appearances free of contradiction, you have done all that can be done towards showing that *q* is free of contradiction, i.e. describes a logically possible state of affairs. In other words, to show that a statement of doubtful standing (e.g. 'There is more than one space') is free of contradiction you have to describe in statements apparently free of contradiction a world in which that statement is true. It is a matter of showing such a world to be conceivable, by describing in detail what it would be like for it to be true in intelligible statements. Thus initially 'There is a man who has two bodies' looks as if it might be logically impossible. But we can go on to describe how it could be true — one subject of experience might be able to control two sets of limbs directly in the way in which we control one set of limbs directly; and acquire knowledge of the world as a result of stimuli impinging on two sets of sense organs. Developing the picture we show in statements apparently free of contradiction what it would be like for the doubtful statement to be true, and thus show it to be logically possible. In this book I shall spell out what it would be like for there to be two spaces, and for physical geometry to be non-Euclidean, and thereby show these to be logical possibilities.

You can show that a statement *q* entails a contradiction, by deducing that contradiction from it. But to do that, you need to know what a statement entails. It is not always obvious what a statement entails. You can show that a statement *q* entails a statement *p*, by showing that the only worlds in which *q* holds of which there seems to be an intelligible description are also worlds in which *p* holds. Of course the fact that a given individual cannot describe a world in which '*q* but not *p*' holds in ways which he finds intelligible does not conclusively prove that *q* entails *p*; but the fact that no one can give such a description is evidence (and the only evidence there can be) of what the speakers of the language are committed to when they utter the sentence which makes the statement that *q*. Thus to show that 'There is a surface which is red and green all over' is logically impossible, you need to show that 'This is a surface which is red all over' entails 'This is not green all over'. The only way to show this entailment is to show that any attempt to describe a surface which is red and green all over ends up with an unintelligible description. In this book I attempt to show, for example, that 'There is a four-dimensional world' is logically impossible by showing that any attempt to describe such a world ends up unintelligibly.

Many writers in the tradition of logical positivism have claimed that a statement can only be a synthetic or factual statement if there could (logically) be some observable evidence which, if observed, could count for or against it. This doctrine is often called verificationism. 'Once upon a time the Earth was red-hot' is, on this doctrine a synthetic statement because it could be confirmed or disconfirmed by geological evidence now available, or indeed because it could have been observed to be true at an earlier time by rational beings situated on other planets. By contrast, positivists claimed, a non-analytic statement such as 'Reality is one' which for logical reasons could not be confirmed or disconfirmed by observation was really logically impossible or in their terminology 'meaningless'. However I know of no good argument for accepting verificationism. A sentence may make a perfectly meaningful statement — the words which occur in it may be meaningful and be put together in accord with normal grammatical rules; and there may be no contradiction involved in the statement or its negation — and yet there may be reason of logic why the statement is not in any way confirmable or disconfirmable by observation. An apparent example is 'There is an uninhabited planet on which a tree dances once a year without there being any evidence of its activity'. (Under evidence I include both physical traces left behind by this tree, and evidence of trees observed to dance in similar circumstances elsewhere.) Clearly there can be no evidence of observation that an event of a certain kind occurs in circumstances where there is no evidence of its occurrence. Nor can there be evidence of observation against it; for it is very unlikely *a priori*, and there is nothing particular which, if observed, would make it any less likely. So the cited statement cannot in any way be confirmed or disconfirmed by observation, and yet there seems no reason apart from verificationist dogma for believing that it is not a meaningful synthetic statement (although of course not one which we have any reason for believing to be true). So mere unconfirmability by itself does not show lack of factual meaning.

However the converse does appear to hold. If there can be evidence of observation which confirms some statement, that statement must make a meaningful factual claim; for if it contained a contradiction nothing could add to the likelihood of its truth. There are certain normal inductive standards according to which evidence would confirm a statement, if that statement is not logically impossible — e.g. the statement is confirmed if it successfully predicts the evidence and is simple (in the way which I spell out on pp. 43ff). If there could be such evidence which apparently confirms a statement then (in the absence of positive grounds for supposing the statement to contain a contradiction) that suggests that it really does confirm the statement and so that the statement is a meaningful factual statement. I shall therefore sometimes argue that where some claim about space and time (e.g. the existence of Absolute Space, which I discuss in Chapter 3) could apparently be confirmed by evidence of observation, that provides reason for believing it to be a meaningful factual statement.

As well as distinguishing between the logically necessary, possible, and impossible, I need a further distinction between the physically necessary, possible and impossible; and the merely practically necessary, possible and impossible. The physically necessary is whatever is necessitated by the laws of nature. The physically impossible is whatever is ruled out by them; the physically possible is whatever is permitted by them. Thus (given the truth of the current physical theory of matter) it is physically necessary that the quantity of matter-and-energy be conserved in any physical interaction. It is physically impossible (given the truth of modern Quantum Theory) for an observer to determine the position (x) and momentum (p_x) of any particle more accurately than within the ranges Δx, Δp_x, where $\Delta x.\Delta p_x \geqslant \dfrac{h}{4\pi}$, h being Planck's constant. Or, to take an example which we shall have to consider later in much detail, it is physically impossible to travel between two points faster than light (c. 300,000 km per sec). It is on the other hand physically possible for an observer to measure the position and momenta of all particles within a room within the ranges Δx, Δp_x $\left(\Delta x.\ \Delta p_x \geqslant \dfrac{h}{4\pi} \right)$, or to travel at more than 200,000 km per sec over some surface.

What is physically impossible is what is ruled out by the permissible ways in which physical objects change their positions and characteristics, or by the permissible ways in which, from whatever arrangement they start, they can be arranged. What is practically impossible is what is ruled out not merely by this but by the actual arrangements of the constituents of the Universe at the time at issue. It is what cannot be done by a man given the way in which the constituents of the Universe behave *and* the way they are at present arranged. Thus, provided with a suitable calculating device or the material, labour, and plans for constructing one, I could find out within the above-mentioned range the positions and momenta of all the particles in this room at some given temporal instant. Provided with a suitable space rocket or the material, labour, and plans for constructing one, I could travel across the Earth at 200,000 km per sec. I have not got these devices nor the material, labour, and plans for constructing them and so the doing of the aforementioned things is not practically possible — at any rate, for the immediate future. What is not practically possible now may become practically possible if, by human effort or otherwise, the constituents of the Universe become differently arranged. If, given their present arrangement and the laws of nature governing their change of arrangement, they could never adopt a future arrangement permitting the doing of some thing, the doing of that thing will never be practically possible. Yet if, whatever the present arrangement of the constituents of the Universe, a given effect could never be produced, the producing of it is physically impossible. What a human agent can now effect given the present arrangement of the constituents of the Universe is now practically possible. What is practically necessary is what, if anything, a human agent in these circumstances must effect.

These distinctions I have made only loosely, as my work is not devoted primarily to making such distinctions but to using them to clarify the characteristics of Space and Time. But I hope that I have made them with sufficient clarity for my use of them to provide subsequent illumination.

BIBLIOGRAPHY

[1] I. Kant, *Critique of Pure Reason*, introduction.
[2] W. V. O. Quine, *From a Logical Point of View*, Cambridge, Mass., 1953, ch. 2, 'Two Dogmas of Empiricism'.

[3] H. P. Grice and P. F. Strawson, 'In Defence of a Dogma', *Philosophical Review*, 1956, 165, 141–58.

[4] J. F. Bennett, 'Analytic–Synthetic', *Proceedings of the Aristotelian Society*, 1958–9, 59, 163–88.

For a fuller exposition and justification of the main distinctions and claims made in this Introduction see;

[5] Richard Swinburne, *The Coherence of Theism*, Oxford, 1977, chs 2–3.

1 Place and Matter

At every instant of time every material object which exists at that instant occupies some place, and wherever any material object is or, it is logically possible, could be, is a place. A place in the literal sense is wherever a material object is, or, it is logically possible, could be. Hence places have some volume.[1] Let us call the place of an object, described as precisely as is logically possible by specifying its boundaries, the primary place of the object. The primary place of a material object will in consequence be the volume enclosed by a surface which fitted snugly round the material object, completely enclosing it.[2] If my shaving-soap stick fits snugly into my shaving-soap case, the most precise description which can be given of where my shaving-soap stick is is that it is in my shaving-soap case. When an object which has a place is not the kind of object which can be enclosed by a snugly fitting surface (for examples see pp. 16f) then other considerations are relevant to determining the most precise description which can be given of where it is and so its primary place.

Where an object is can, however, be described more loosely than by identifying its primary place, and this in two ways. First it can be described by identifying any place enclosing the primary place. For any place which encloses the primary place of an object is the place of the object. Since my shaving-soap case is in the bathroom, my shaving soap is in the bathroom. Since the bathroom is in my house, the shaving soap is in my house, and so on. It would be misleading to say that the shaving soap is both in the case and in the bathroom and in the house, for this would suggest that it was in several different places at once. The shaving soap is only in one place at the temporal instant in question, but that place can be more or less precisely described.

---

[1] Two- or one-dimensional entities like patches of light or lines are not normally said to occupy places but to have positions. Such uses of 'place' as 'the place of an argument in the thesis' or 'the place of religion in the home' are clearly metaphorical.
[2] Aristotle ([1], Chapter 2) regarded the enclosing surface as defining the primary place of a material object.

Alternatively where an object is may be described by identifying smaller places or points enclosed within the object. My field may be said to lie at the intersection of the 55th parallel of latitude north and the first meridian of longitude west. What this means is that the place of my field is a region of unspecified volume surrounding the point formed by the intersection of those two lines. Both identifying a place which contains among other distinct places the primary place, and identifying a smaller place or point enclosed by the object, give less precise descriptions of where an object is than the most precise description which can be given.

A place is identified by describing its spatial relations to material objects forming a frame of reference, and which are for this purpose regarded as fixed. Any set of material objects which over a period of time retain the same spatial relations among themselves may form a frame of reference and will form such a frame if they are used for locating places and objects. Other material objects which over that period of time retain the same spatial relations with certain material objects forming a frame of reference form the same frame as they do.

An object (or place) is spatially related to another object (or place) if it is in some direction at some distance from it. The spatial relations between objects (and places) are their distance and — given the applicability of the concept (see Chapter 4) — direction from each other. Two objects retain the same spatial relations if they remain at the same distance in the same direction from each other (it is logically necessary that whatever spatial relations two places defined by reference to the same frame have to each other at one temporal instant they have at all instants). Thus a group of ships at sea keeping station may form a frame of reference since they retain the same spatial relations to each other. Any rigid body may form a frame of reference since it consists of parts which are material objects, the spatial relations of which to each other do not change. Thus the Earth or a ship can form a frame of reference. One may say where a ship is by saying on what part of the Earth it is situated, or where a wreck is by saying how far it is in what direction from our ship. To say that an object is in the same place as, or is in a different place from, what it was before is to say that its spatial relations to the objects of the frame have not, or have, altered. To say that an object has remained still over a period is usually to say that its parts have not changed their places over that period. The parts change their places either if the object as a whole changes its place or if they merely interchange their places (in the case of a rigid body, if the parts interchange their places, without the rigid body changing its place, the body rotates).

Usually when we talk of objects staying in the same place or moving we do not explicitly mention the reference frame, as it is obvious which frame we are assuming. Thus if we are travelling in a car, and discussing whether the picnic basket is where I put it a quarter of an hour ago, the frame of reference assumed is the car. If I put it on the back shelf and it is still on the back shelf, then it has not moved. To say this this is not to deny that it has moved relative to the Earth, since it is obvious in this case that the frame of reference is not the Earth. The most usual frame of reference is the Earth. To say where something is and that it has or has not moved is, more often than not, to locate it relative to parts of the Earth and to say that it has or has not changed its terrestrial latitude and longitude. But when we are talking of the movements of Mars or Venus it is the Sun or other stars which form the reference frame. Sometimes there is genuine doubt what is the frame of reference and then we have to mention it explicitly, if we are to make our meaning clear. If we are on a ship in convoy and someone says that we have not moved position, we may need to ask him whether he means relative to the other ships or to the land.

It is a consequence of these points that only given a frame of reference, can we reidentify places. We can only say whether or not something is in the same place as it was before if we specify or take for granted the reference frame. What is the same place at a later temporal instant as a specified place given one frame (e.g. the travelling car) will not necessarily be the same place as that place given another frame (e.g. the Earth). However, for one given instant one can sensibly refer to the same place using different frames of reference. I can refer to the place of a ship at midday on Thursday using either the Earth or the convoy to which the ship belongs as the frame of reference. The place identified by its distance and direction from the objects of one frame is said to be the same as the place identified by its distance and direction from the objects of the other frame because the same objects are found at or in the immediate vicinity of each of the places.

Let us now consider the occupants of Space, the things which can properly be said to have a place. Anything which can properly be said to have a place I will call a spatial thing. Spatial things are many and various. They include tables and chairs, rainbows and explosions, quantities of gas, fields of force, quantities of cosmic radiation, people, sounds, and smells. Spatial things may be divided into what I shall call material objects, mere physical objects, contingently dependent occupants of Space, and logically dependent occupants of Space. We shall consider each in turn.

The standard examples of material objects are the solid things which resist our pressure and which we can lift up and push around. Rocks, tables and chairs, people and houses, clothes, cars and plates are all standard examples of material objects. Anything which is composed of such things, in the sense that its spatial parts are such things, is itself also a material object. Thus the Earth and the other planets are said to be material objects because they are composed of things like lumps of rock. Likewise anything of which the standard examples of material objects are composed, in the sense that it is one of the spatial parts of such an object, is itself also a material object. Molecules of which solids are composed are material objects because tables and chairs, etc. are formed from molecules joined together. Further, anything which by occupying a place thereby as a matter of logical necessity excludes other material objects from that place is itself a material object. What is meant by this is not that other material objects cannot get into the place, but that they cannot get into the place without at the same time excluding the other object from it. Anthony Quinton has called this property of material objects the property of logical impenetrability, and has neatly distinguished it from hardness. He writes: 'Nothing can be where the impenetrable is; less hard things cannot get into the place where the hard is' ([5], p. 341). The impenetrability is said to be logical because if anything did occupy the same place at the same instant of time as a material object we would for that reason refuse it the title of material object. A magnetic field may occupy the same place at the same instant as a chair, a sound may occupy the same place at the same instant as a volume of air — but this means that magnetic fields and sounds cannot be material objects. It is in virtue of this criterion of impenetrability that the molecules which compose a gas or a liquid are said to be material objects. If they occupy a place, no other material object can do so at the same time. If there is a volume of air in a cylinder, you cannot keep a solid iron piston there at the same time.

We have thus started with standard cases of material objects and extended the scope of the term by means of the notions of composition and logical impenetrability, thereby fully delineating the class of things commonly called material objects. In so doing we have taken for granted the often urged claim that the following proposition (A) is a logically necessary truth:

(A) No two material objects can be in the same place at the same temporal instant.

This claim and the remarks of the last page must however be qualified in an important way. We are to understand by 'place' primary place. My toothbrush and the shaving soap can both be in the bathroom at the same instant of time, but the toothbrush cannot have the shaving-soap case as its primary place if the shaving soap does. Clearly too a material object can be in the primary place of another material object. A speck of dust can be in the shaving-soap case as well as the shaving soap. But the speck of dust does not have the shaving-soap case as its primary place. The primary place of the speck of dust is a small region within the shaving-soap case.

But could there not be two material objects which completely interpenetrated each other and had the same outer surface? No. Because if two things interpenetrated each other so thoroughly that they could not be separated, they would not be termed material objects. If you mix some strawberries and a lump of cream, all material objects, to make strawberry ice-cream, neither the strawberries nor the lump of cream exist as such any longer. The small particles of strawberry and the small particles of cream are material objects — but of them it is true that they do not have the same primary place as other small particles. If two objects can be separated from each other, then they will have different enclosing surfaces. Anyone who marks off where they are will mark off different regions of Space. I conclude that the following amended version of proposition (A) is a logically necessary truth.

(A') No two material objects can have the same primary place at the same temporal instant.

Material objects, as we have distinguished them, are thus objects possessing the 'primary qualities' to which the philosophers and scientists of the seventeenth century drew our attention. The seventeenth-century philosopher John Locke defined the primary qualities of a body as 'such as are utterly inseparable from the body, in what state soever it be.'[1] These qualities are 'solidity, extension, figure, motion or rest, and number'.[1] Locke seems to understand by solidity 'logical impenetrability' in the sense above defined.[2]

In order that material objects may be recognised, they must have, as well as all the primary qualities, one or more secondary qualities (such as

---

[1] John Locke, *An Essay Concerning Human Understanding*, ii. 8. 9.
[2] See, e.g., op. cit. ii. 4. 4.

colour, taste, smell, or tactual hardness)[1] or powers to produce discernible effects. Only if we can feel the edges of them or see their shape or observe their effects in some way can we recognise the presence of material objects. Most medium-sized material objects we can recognise either visually or tactually. Other material objects we recognise by their effects — we learn about the existence and arrangement of molecules in a chemical substance by observing the behaviour of the substance.[2] Solids are those material objects which do not change their shape readily when subjected to deforming influences.

Material objects are the most important occupants of Space. Indeed, as Strawson [2] has shown, it is only by referring to them that we are able to talk about other things — I identify a certain pain by referring to the person, the material object, whose the pain is; I identify a colour by saying which material objects it characterises, and so on. But, as we have noted, material objects are not the only spatial things,[3] and we must now discuss the other three kinds of spatial thing. Any spatial thing of any other kind must, like a material object, have the primary qualities of 'extension, figure, motion or rest, and number', for if it did not have these properties it could not occupy space. For if an object occupies space it must have a place, and either change its place or stay still, and that place must have some size, and the object must be one in number. It can only differ and will differ from a material object by lacking the primary quality of 'solidity' or logical impenetrability.

Any spatial thing which is not a material object but which has an existence independent of material objects I shall term a mere physical object. By the object having an existence independent of material objects, I mean that statements about it cannot be reduced to statements about material objects, and also that it is not physically necessary for its present existence that there be material objects of certain kinds behaving in certain ways. As with material objects, we learn about mere physical objects by their secondary qualities or by their effects. Quantities of radiation such as a photon (the particle of light) or a beam of radio

[1] Locke of course wanted to say the secondary qualities were 'nothing in the objects themselves, but powers to produce various sensations in us by their primary qualities', op. cit. ii. 8. 10. But this is a further issue which it is not necessary to discuss here.

[2] In an important recent article J. F. Bennett has shown that disagreement about which primary qualities bodies have is incompatible with coherent talk about them, whereas disagreement about the secondary qualities of bodies is perfectly compatible with coherent talk about them. J. F. Bennett, 'Substance, Reality, and Primary Qualities', *American Philosophical Quarterly* (1965), 2, 1–17.

[3] For the rival view that material objects are the only spatial things see [5].

waves are examples of mere physical objects. They are spatial things for we can say where the photon is or talk about the quantity of radiation in interstellar space. Yet photons are not material objects, for the primary place of a photon may be the primary place of another photon or of a material object, such as an atom of carbon. The most precise description which can be given of where an object is describes its primary place. Yet if I specify as precisely as possible the region within which a photon lies, another photon or material object may have that region as its primary place. This is no mere physical limitation on our ability to measure where photons are, for an object would not be a photon if we could shut it up within a surface which fitted snugly round it and thereby exclude all material objects. A photon is not an object of that kind. Further, radiation has an existence independent of matter, for you could have a universe filled solely with radiation.

Material objects and mere physical objects I group together as distinct occupants of space which have an existence independently of each other in a class of physical objects. All the physical objects to which science has drawn our attention behave in fact remarkably like material objects. Quantities of radiation have a certain energy and velocity and certain types of material object can prevent radiation from entering a region of Space. Matter can be transformed into radiation, and radiation into matter. But it is possible to imagine physical objects with properties much more distinct from those of material objects. Ghosts, if such there were, would be spatial things, not material objects but having an independent existence. Hence they would be mere physical objects. Yet it might not be possible for them to be transformed into matter or matter into them; and they might have very few properties in common with material objects. However, we have little good reason for believing in the existence of ghosts, and the only physical objects in the existence of which we do have good reason for believing are the quasi-material particles of science such as photons and neutrinoes and collections thereof.

Now whatever occupies space must be a material object or have an existence independent of material objects and so be a mere physical object, or depend on physical objects for its existence. The dependence can be either logical or contingent. Hence of logical necessity there can only be these four sorts of spatial things — material objects, mere physical objects, logically dependent occupants of space, and contingently dependent occupants of space.

Logically dependent occupants of Space are spatial things other than physical objects which depend on physical objects in a logically necessary way. That is to say, talk about such logically dependent things is just talk about the way in which physical objects do or would behave. Thus a gravitational field (in the sense in which this is understood in Newtonian physics) is a logically dependent occupant of Space. It is a spatial thing for we can say where there is a gravitational field of such and such a strength and direction. But to say that there is a gravitational field of a certain sort in some place is simply to say that physical objects in that place are subject to a certain force proportional to their mass.

Contingently dependent occupants of Space are spatial things other than physical objects, which depend for their existence on physical objects, as a matter of physical, not logical necessity. Examples are smells and sounds. These have places and so are spatial things. Obviously they are not physical objects. They would not be there, were there not physical objects which gave rise to them. There would not be a smell of beef in the kitchen were there not small particles of beef floating about in the kitchen air. Nor would there be sounds, were there not physical objects in the form of air molecules vibrating in certain ways. Yet this dependence is a contingent and not a logical matter. Smells would be smells and sounds sounds if they could have an existence independent of material or other physical objects. But then they would themselves be physical objects and not contingently dependent occupants of Space.

In later chapters we shall be considering questions about the size and age of the Universe. In discussion of such questions, I take 'the Universe' to mean all the physical objects that there are, spatially related to the Earth.[1] Physical objects are the occupants of Space which have an existence independent of each other and which, as a matter of fact, behave in ways very similar to the hard solid obvious constituents of the Universe like tables and chairs. To understand by 'the Universe' all the physical objects that there are, spatially related to the Earth, seems to bring out the way in which the term has functioned in cosmological discussion.[2] If there are physical objects not spatially related to the Earth — the logical possibility of which we shall consider in the next chapter — it would be most natural to say that they belong to another universe.

---

[1] The significance of this latter clause will become clear in Chapter 2.
[2] An alternative suggested definition of 'the Universe' will be examined and rejected in Chapter 12 (see p. 212). Kant's objections to the legitimacy of all talk about the Universe and its properties will be examined and rejected in Chapters 14 and 15.

Of spatial things, the following proposition is often urged as a logically necessary truth.

(B) No spatial thing can be in two different places at the same temporal instant.

As with (A), we must understand by 'place' 'primary place', since, as we have noted, my shaving soap can be both in the shaving-soap case and in the bathroom. An object can be in two places at once if one place encloses the other place, but then at most one of those places will be the primary place of the object. An object can also be in two places at once, if a part of the object is in one place and a part in another. The Soviet third army will be both in Uzbekistan and in Armenia, if half of it is in Uzbekistan and half of it in Armenia. But then neither Uzbekistan nor Armenia is the primary place of the Soviet third army. The primary place of that army is a region including a part of Armenia and a part of Uzbekistan. The army cannot be in Moscow at the same temporal instant as having that region as its primary place. So (B') below is a logically necessary truth.

(B') No spatial thing can have two different primary places at the same temporal instant.

We classify chunks of matter as material objects of various kinds, according to the similarities between them which strike us naturally or have practical importance for us — as tables, trees, houses, plants, fences, stars and mountains. (In the terminology of [4] 'table', 'tree', etc. are different substance-sortals. For general principles governing the kind of substance-sortals there can be, see [9].) A material object of one kind K (e.g. a fence) may include as its parts material objects of a different kind (e.g. planks). Sometimes we may have two different ways of classifying the same chunk of matter — e.g. we may classify a certain fence either as a fence or as an assemblage of planks. We could call this a case of two different material objects occupying the same place (and so as constituting a counter-example to Principle (A)); but it seems more natural to describe it as a case of the same material object being differently classifiable. How we classify a material object will affect its criteria of identity, the criteria for saying that a later object is the same material object as the original object. If the planks of the fence are gradually replaced by other planks we still have the same fence but not the same assemblage of planks.

For the continuing identity of material objects over time, in general two conditions of spatio-temporal continuity need to be satisfied; one concerning the continuity of the whole and the other concerning the continuity of its parts. The first is that two material objects of kind K, M at time t and M' at time t' must be spatio-temporally continuous if they are to be the same material object. By this is meant that there must be a material object M'' of kind K similar in its properties to both M and M' at every temporal instant t'' between t and t', such that each M'' at each t'' occupies a place contiguous with the place occupied by the M'' at the prior and succeeding instants, however precisely temporal instants are identified, the series beginning with M at t and ending with M' at t'. Thus for this desk at which I am writing to be the same desk as the desk at which I was writing yesterday in this room, there has to be a spatio-temporal chain of similar desks joining them. If this desk just appeared from nowhere, or was linked spatio-temporally with a desk in another room yesterday, then this desk would not be the desk at which I was writing yesterday.

Some further condition about spatio-temporal continuity of parts also needs to be satisfied; but the form of this condition will vary with the kind of object in question. For many objects the condition will be that most parts of equal volume of the M'' at each t'' must also occupy places contiguous with a place occupied by a part of the M'' at the prior and succeeding instants. Thus if all of the parts of my bicycle except the front wheel were at one instant replaced with new parts, we would not say that the new bicycle was the same as the old one. But we probably would say this if the replacement was gradual — one wheel being replaced one month, the brakes being replaced the next month, and so on. In the latter case at each instant there would be spatio-temporal continuity of most of the parts. For objects of certain kinds the spatio-temporal continuity of parts need concern only some central controlling part. Thus for animals it is natural to say that we have the same animal if it has the same brain even if it has new legs, arms etc., i.e. if there is spatio-temporal continuity of the brain, but not of legs and arms. For a few objects what matters is spatio-temporal continuity of most of the parts with an original object. In these cases total replacement of parts, however gradual, would mean that the original object had ceased to exist. The condition of spatio-temporal continuity of parts takes this form for objects of great historical interest. If you gradually replace most of the stones of the Parthenon with new stones, the original Parthenon no longer exists.

The joint satisfaction of the two conditions of spatio-temporal continuity, in the form appropriate to the kind of object in question, will guarantee that only one object M' at t' is the same object as M at t. Only one bicycle at t' can have originated from a given bicycle at t if only gradual replacement of parts is allowed. But if any bicycle is the same bicycle as the earlier bicycle if it has a part in common with it, there could be no end of later bicycles identical with the earlier bicycle. In that case principle (B') above would be violated; the earlier bicycle would be simultaneously in two different primary places. Sometimes the first condition may be relaxed and satisfaction of the second condition be deemed sufficient to constitute identity over time. This possibility will arise if an object such as a bicycle at t is taken to bits and then put together again at t'. The first condition will not be satisfied, because there is a period between t and t' at which no bicycle existed. The two bicycles are not joined by one spatio-temporal chain. But each of the parts of the bicycle are so joined. Yet the relaxing of the first condition can lead to the existence of two objects at t' which are rivals for being the same object as the object at t. The planks of the ship of Theseus were gradually removed one by one and replaced with similar planks. The latter were then put together to form a ship. There are then two ships at a later time which have a claim to be the same ship as the earlier ship. One (a) has all the parts of the earlier ship, and the other (b) has preserved continuity as a ship. Both have satisfied a condition of continuity of parts in the first form in which I stated it; though (a) has of course satisfied a much stronger form of this condition. Only (b) however has satisfied the condition of continuity of whole. If we are prepared to drop the condition of spatio-temporal continuity of whole, some arbitrary decision may be required to choose between rival claimants for being the same object as an earlier object. (See [7] for a discussion of these problems of spatio-temporal continuity of parts).

Such then, with qualifications, are the conditions of spatio-temporal continuity which must be satisfied for the identity of a material object over time — at any rate for material objects of almost all kinds. (A possible exception will be suggested in Chapter 2.) But does there not then arise a circularity of definition? For a necessary condition of M and M' being the same material object is that they must have occupied places of certain sorts at certain temporal instants. But places, we have seen, can only be identified by relation to material objects. 'So the identification and distinction of places turn on the identification and distinction of things; and the identification and distinction of things turn, in part, on the identification and distinction of places' ([2], p. 37).

This circularity is not, however, vicious. A vicious circularity of definition only arises where one expression is defined in terms of another and where one cannot even make a provisional judgement of the applicability of one expression until one has correctly (and knowingly so) applied the other expression, and conversely. For in such circumstances one could never understand the one expression without understanding the other expression, and conversely; and so one could never get started. But our circularity is not a vicious one, for we can make a provisional judgement that M and M' are the same material object on the basis of very precise satisfaction of a different criterion, viz. their having almost precisely similar properties apart from those of spatial relation. We take the detailed similarity between a present object and a past object as evidence that the two objects are the same object. We take two tables with scratches in the same place, two books with ink blobs on the same pages etc. as the same object, in the absence of positive counter-evidence, e.g. that they are not spatio-temporally connected or that there are two or more virtually qualitatively identical objects of that kind in the world. This criterion of similarity seems to be an inductive principle, a principle of what is evidence for what in the world, which follows from the very general inductive principle of simplicity. Among hypotheses which lead us to expect the data of observation, the simplest is that most likely to be true. It is simpler in explanation of our observations to suppose that there is only one entity than that there are two more or less qualitatively identical ones — and so in the absence of evidence better explained by the hypothesis of two entities, it is more probable that there is only one.

We retract any one provisional judgement made by the criterion of similarity when we have evidence that the conditions of spatio-temporal continuity do not apply, if we assume that most of the other provisional judgements made on the basis of the criterion of similarity are correct. When I wake up and look around I judge that the room in which I am is my bedroom, the room in which I went to sleep. My judgement is made solely on the basis of its appearance. I do not bother to find out whether or not it has been spatio-temporally continuous with my bedroom yesterday. But suppose I go downstairs and the rest of the house looks very different from what my house looked like yesterday, then either the bedroom was not really my bedroom, or the bedroom has moved, or the rest of the house has changed overnight. Once there is serious doubt about whether the bedroom in which I woke up was my bedroom (that is the room in which I went to sleep yesterday), despite the satisfaction of the criterion of similarity, I must investigate whether or not the conditions of spatio-temporal continuity are satisfied for the two bedrooms. To do this I have provisionally to reidentify other objects by the criterion of similarity. I look, that is, for landmarks in order to find out whether the bedroom in which I woke up was in the same place as the bedroom in which I went to sleep. If it is not in the same place, then unless I can produce evidence of spatio-temporal continuity of the two by giving a plausible account of how and why my bedroom changed its position (something of a kind which I know does not normally happen), the conditions of spatio-temporal continuity have been shown not to be satisfied for the two bedrooms. I walk into the road and find a few hundred yards away a house very similar to my own, and all the objects — trees etc. — surrounding it very similar to the objects surrounding my house. I therefore provisionally identify that house as my house using the criterion of similarity alone. It then follows that either my bedroom was transported bodily during the night or that the room in which I woke up was not my bedroom. If there is no evidence making plausible the former hypothesis, I must adopt the latter. Despite appearances the room in which I woke up was not my bedroom. My ability to make any well substantiated judgements of reidentification of material objects thus depends on my ability to make some provisional judgements on the basis of the criterion of similarity alone. But if I could not provisionally reidentify any objects without proof of the satisfaction of the conditions of spatio-temporal continuity, I could never get started.

What this brings out is that I can seriously doubt whether I have correctly reidentified some material objects and my doubts can be settled, but if I seriously doubt whether I have reidentified *any* material object correctly, then this doubt cannot be settled. However, though we can often provisionally reidentify a material object without reidentifying places, we can never even provisionally reidentify a place without reidentifying material objects. Places have no independently recognisable qualities. They are what they are by virtue of the material objects which surround or occupy them. Granted that I have some ability to reidentify material objects, I can doubt of all places whether I have correctly reidentified them and my doubts can be resolved. There is thus a sense in which 'the identification and distinction of places turn on the identification and distinction of things' without 'the identification and distinction of things' turning 'on the identification and distinction of places'. If there are some objects as I shall be arguing in Chapter 2, for the identity of which spatio-temporal continuity is not a necessary condition, that further emphasises the asymmetry of dependence between material objects and places.

There is a second difficulty which arises in this connection. If in order to show that M and M' are the same material object at different temporal instants we have — at any rate sometimes — to show their spatio-temporal continuity, do we not presuppose a frame of reference in so doing? For the definition (p. 19) of spatio-temporal continuity used the notion of a place being contiguous to another place occupied at a prior temporal instant by a certain object, that is, to a place which is the same place as that occupied previously by a certain object. But what is the same place as a place at a different instant depends, as we have seen, on the frame of reference which we choose. But then how can we show spatio-temporal continuity without presupposing a frame, and in that case is not what counts as the 'same material object' dependent on the frame we choose, just as is what counts as the 'same place' (as was first suggested by D. M. Armstrong [3])? To an important extent re-identification of material objects is not dependent on the frame chosen. For if M and M' are spatio-temporally continuous when places are judged by a frame F, they will be spatio-temporally continous when they are judged by any frame F' moving continuously with some (possibly zero) uniform or non-uniform velocity relative to F. If the conditions of spatio-temporal continuity hold when the places are judged relative to the Earth, they will hold when the places are judged relative to any space rocket moving continuously relative to the Earth. A thing moves continuously relative to a frame F, if, after leaving any place, it occupies at the next instant of time, however precisely temporal instants are distinguished, a place which, judged from frame F, is contiguous to the former place. A group of frames, measurements relative to which yield the same judgements of spatio-temporal continuity and thus the same judgements of 'same material object', I shall term an MO-invariant group. The MO-invariant group which includes the Earth will be called the E-group, and any member of it an E-frame. A frame would be a non-E frame if after leaving a place p, as judged from the Earth, at the next distinguishable instant, it occupied a place p' not contiguous with p, but at some distance from p, as judged from the Earth; and so got from p to p' without passing through any places lying between them as judged from the Earth.

The conditions of spatio-temporal continuity for the identity of material objects are thus elliptical, because spatio-temporal continuity depends on the frame of reference and those conditions do not specify the frame. Clearly M and M' will be the same object if the conditions are satisfied, when spatio-temporal continuity is estimated relative to an E-frame. For in all the judgements which are paradigm examples of correct judgements about identity of material objects — such as that this is the same desk as the one on which I was writing two minutes ago — spatio-temporal continuity is estimated relative to E-frames. However it does not seem to be necessary that spatio-temporal continuity should be assessed relative to an E-frame. For surely there could be persons who interacted with their surroundings as we do, when those surroundings seemed to them to have just the sort of stability that ours do, without the constituents of those surroundings which seemed to the inhabitants to be unchanging, being spatio-temporally continuous relative to the Earth. The series of objects which they were inclined to call one continuing tree might be a series of qualitatively similar objects which lacked spatio-temporal continuity relative to the Earth, and yet preserved it relative to other series of objects which they were also inclined to call continuing things. And the series of objects which each person was inclined to call his own body might also exhibit spatio-temporal continuity relative to these other series, and yet not relative to the Earth. (The objects of each series might form a series of objects spatially related to the Earth — e.g. now in this part of the sky, now in that, or in the way to be suggested in Chapter 2 they might have no spatial relation to the Earth at all.) Surely the series of objects of those surroundings have as much right to be called continuous material objects as do the series of objects which form our paradigm examples of material objects.

I suggest therefore that all talk about spatio-temporal continuity and so all talk about the identity of material objects involves implicit reference to an MO-invariant group. For us on Earth this is always the E-group. But it is possible to identify material objects relative to another group, and there might be people whose surroundings had the sort of stability which made it natural for them to make their judgements relative to a different MO-invariant group. They would make provisional judgements to identity on the basis of the criterion of similarity. There would be enough such judgements to yield a stable frame of reference (which was not an E-frame), by means of which they could check individual such judgements to see if they satisfied the conditions of spatio-temporal continuity relative to their frame.

BIBLIOGRAPHY

[1] Aristotle, *Physics*, book iv, ch. 1–5.
[2] P. F. Strawson, *Individuals*, London, 1959, ch. 1.

[3] D. M. Armstrong, 'Absolute and Relative Motion', *Mind*, 1963, 73, 209–23.

[4] David Wiggins, *Identity and Spatio-Temporal Continuity*, Oxford, 1967.

[5] Anthony Quinton, 'Matter and Space', *Mind*, 1964, 74, 332–52.

[6] David Sanford, 'Volume and Solidity', *Australasian Journal of Philosophy*, 1967, 45, 329–40.

[7] Robert C. Coburn 'Identity and Spatio-Temporal Continuity', in Milton K. Munitz, (ed.), *Identity and Individuation*, New York, 1971.

[8] Brian Smart 'How to Reidentify the Ship of Theseus', *Analysis*, 1971–2, 32, 145–8; and his reply to two criticisms in intermediate issues of *Analysis*, 'The Ship of Theseus, the Parthenon, and Disassembled Objects', *Analysis*, 1973–4, 34, 24–7.

[9] Eli Hirsch, 'Physical Identity', *Philosophical Review*, 1976, 85, 357–89.

2 Spaces

All places (by whatever frame identified) spatially related to each other taken together constitute a space. A space, in the literal sense of the term, is that in which material objects are situated, and move or remain still. They change their place by moving through space.

Space is of logical necessity unbounded. This is to say that every region of space is of logical necessity surrounded on all sides by other regions of space. The denial of this is the assertion that it is logically possible that there be a boundary to space. A boundary to space would be, as it were, a wall barring further progress. Yet it must be that behind this wall either there is some object, in which case that object would occupy space; or there is no object, that is there is unoccupied space. It does not seem possible to give a coherent description of any third possibility. Further, between any two non-contiguous places, A and B, spatially related to each other, there lies another place. For if you set off on the shortest path from A to B (and since they are spatially related, there is a path between them) either there will be an obstacle in the way or there will not. If there is an obstacle, then it occupies a place. If there is no obstacle, then there is an unoccupied place between them. There must be, for otherwise you could not pass along the path.

Scientists and philosophers sometimes dispute about whether space is discrete, dense, or continuous. These are basically claims about the number of points, or smallest bits of space, which lie on a line between any two non-contiguous places. To say that space is discrete (or granular or atomic) is to say that there are only a finite number of points between any two non-contiguous places. To say that space is dense (but not continuous) is to say that there are an infinite number of points between such places, but a denumerably infinite number. A denumerable infinity is the number of the rational numbers, that is numbers which can be expressed as fractions with integers as numerators and denominators, e.g. $1/4$, $8\frac{1}{4}$, $2\frac{1}{8}$, $\frac{221}{227}$. To say that space is continuous is also to say that there are an infinite numbers of points between non-contiguous places, but the larger infinity of the continuum. This is the number of the class of rational and irrational numbers; that is, including such numbers as $\sqrt{2}$,

$\sqrt[3]{2/5}$, π and π^2 which cannot be expressed as fractions with integers as numerators and denominators. So much is clear, but what it is for there to be this or that number of points between non-contiguous places is quite unclear, and left unclear by most writers on the subject, and it needs to be made clear before these claims can be assessed.

What would it be like for space to be discrete? If there are only a finite number of points between A and B, these points will be spatial atoms which take up a certain finite volume. But, then, with respect to a given atom one can talk about the half of it closest to A, and one can say how big that half is, and how far it is from A. Is not this half a volume of space? So how can the atom be the smallest bit of space? If space is dense or continuous, points will have no finite volume, and so this difficulty does not arise. But what are we saying when we say that the number of points between A and B is dense rather than continuous? Since between points are distances, what the claim would seem to amount to is the claim that there can be distances only of n units of some kind, where n is a rational number. If space is continuous, there can be distances of n units, where n is an irrational number. But what is it for there *not* to *be* a distance of $\sqrt{2}$ units? Clearly one can always talk and make calculations about intervals of that length in any large space, and about how bodies would behave if they occupied such an interval, and so on. I suggest that the only sense that can be given to these various claims is in terms of what sizes of material object there can be and what distances they can cover. To say that space is discrete is to say that there is some finite length k, such that it is physically possible that material objects be of length k, $2k$ etc. units but not $1\frac{1}{2}k$ units. That is a comprehensible claim, and one for which there could be evidence (e.g. finding that all observed material objects had such lengths for some finite length k within very precise limits of accuracy). I do not know of good scientific evidence for or against this claim. Likewise the claim that space is dense (but not continuous) could be a claim about what lengths it is physically possible for bodies to have, that their length can only be a rational number of units of some kind, but that they cannot be of length $\sqrt{2}$ units, or π units. Conversely, the claim that space is continuous would be the claim that bodies can be of any length any real number (e.g. $\sqrt{2}$ or π) of units of some kind. These claims seem coherent enough (once the unit of length is specified). But it also seems to be the case that no evidence could ever confirm or disconfirm the claim that space is dense but not continuous, or vice versa. For any evidence about the measured size of objects could be explained equally well on either theory. For you can always find a rational number which lies within any finite interval you choose, however small, of a given irrational number. Yet measurement can only be made to a certain finite degree of accuracy; and so any measured interval can be represented as of length a rational number of units of some kind within that limit of accuracy. Given that the rival theories are equally simple (see pp. 43ff); they would always be equally well supported by evidence. (See [4].)

A place *A*, we saw in the last chapter, is spatially related to another place *B* if *A* is at some distance from *B*. What this means is that measuring rods could be laid on a path between *A* and *B* by a being who took sufficient time and was powerful enough to overcome all forces opposing the journey. It may of course be practically impossible for a man to be able to construct a machine of sufficient power to overcome opposing forces or to live long enough to make the journey. But in saying that two places are spatially related I am saying that if there are no such reasons, then measuring rods can be laid between one and the other. This is a logical truth about the meaning of 'spatially related'.

The relation of being spatially related is clearly symmetrical and transitive. If *A* is spatially related to *B*, then *B* is spatially related to *A*. If measuring rods can be laid between *A* and *B* under certain conditions, they can be laid between *B* and *A* under those conditions, for to achieve the one task is to achieve the other task. If *A* is spatially related to *B*, and *B* is spatially related to *C*, then *A* is spatially related to *C*. For if measuring rods can be laid under certain conditions between *A* and *B* and between *B* and *C*, they can be laid between *A* and *C* under those conditions, if need be via *B*.

Now is it true of logical necessity that all places are spatially related to each other? If so, we can only talk of the one space. But if it is logically possible that there can be places not spatially related, then we can talk of different spaces. Kant thought that the former was an *a priori* truth. 'We can only represent to ourselves one space, and if we talk of diverse spaces, we mean thereby only parts of one and the same unique space.' ([1], B.39) 'If . . . I . . . say that all things, as outer appearances, are side by side in space, the rule is valid universally and without limitation' ([1], B.43). In an important article, 'Spaces and Times' [2], Quinton has argued that Kant is wrong, and that it is not an *a priori* truth and so not a logically necessary truth that all places are spatially related to each other.

Quinton makes his point by giving us a Two-Space myth; that is, he describes in statements apparently free of contradiction, a world in which there are two spaces; and, as he describes it, a world in which an inhabitant would have evidence of observation which confirmed the claim that there were two spaces. Quinton asks us to suppose that our dream-life undergoes a remarkable change:

> Suppose that on going to bed at home and falling asleep you found yourself to all appearances waking up in a hut raised on poles at the edge of a lake. A dusky woman, whom you realise to be your wife, tells you to go out and catch some fish. The dream continues with the apparent length of an ordinary human day, replete with an appropriate and causally coherent variety of tropical incident. At last you climb up the rope ladder to your hut and fall asleep. At once you find yourself awaking at home to the world of normal responsibilities and expectations. The next night life by the side of the tropical lake continues in a coherent and natural way from the point at which it left off. And so it goes on. Injuries given in England leave scars in England, insults given at the lakeside complicate lakeside personal relations. . . . Now if this whole state of affairs came about it would not be very unreasonable to say that we lived in two worlds ([2], pp. 141f).

In this myth, Quinton argues, our life in the one set of surroundings is just as coherent as that in the other, and so there are no grounds for saying that one of the lives is only a dream. There are other people by the lake and in our ordinary surroundings, and to anyone who urges that one group of people does not really exist, that we only dream about them, the answer must be made that whatever grounds there are for saying that, there are the same grounds for saying it about the other people. There are, however, on Quinton's myth, no grounds for saying that the objects of one set of surroundings are spatially related to those of the other, for however much exploration we do in one life, we can never find the objects of the other.

Now is Quinton's myth coherent? Does it describe, without contradiction, a world in which there are two spaces, and a world in which an inhabitant would have grounds for believing that there are two spaces. I shall argue that, with certain amendments to the myth, Quinton is successful in showing what he wishes to show. But, first, I must deal with four objections to Quinton's argument. Some are objections to the claim that Quinton has given a coherent description of a world of two spaces, and some are objections to the claim that an inhabitant would have grounds to believe that he lived in a two-space world.

The first objection is that there is an inconsistency in Quinton's account as stated in that while a man is in England, living the life of the

English day, his body, according to Quinton, lies asleep by the lakeside[1] — and presumably, conversely. The man is only awake in one place at one temporal instant, but asleep and awake, he is to be found in two places at one temporal instant. And this violates the principle which we have noted (p. 18) to be a logically necessary truth, that no spatial thing can have two different primary places at the same temporal instant. We might attempt to avoid this difficulty by urging that the true man is the man's mind or soul and that that was only in one place at one time, and that the man's mind was connected with two different bodies. But all this would need a lot of philosophical justification to make coherent, and there is no need to take this route to meet the objection. We need only amend the myth so that soon after people who have these odd experiences go to sleep, they just vanish from their beds, and soon after they reappear, they wake up. Then it will not be the case that they are in two places at one temporal instant.

The second objection is that although a man who had the sort of experiences which Quinton describes would have no grounds for saying that life in one set of surroundings was real while life in the other set of surroundings was only a dream, he might have good grounds for doubting the reality of both lives and for believing that he was subject to perpetual hallucinations. His experiences might seem so mysterious and disordered that he reasonably doubted whether either sort of experience was properly described as genuine. This might, I think, be the case if only one man had this sort of experience and all the other people of both his environments reported to him (or appeared to report to him) that they did not. For then in each set of surroundings when he told people of his other life, they would not believe him — since they would have no other evidence than his word for crediting what he said. Hence throughout his life everybody he met would doubt the truth of half of his reports of previous events. They would say that he 'only dreamt it'. And each of his reports would be doubted on half the occasions on which it was made. In such circumstances the man might reasonably hold that he had no grounds for believing any of his own reports of past events, that his whole life, and not merely part of it, was a dream. At that point he would presumably believe himself to be unable to wake up, and so would become demented.

The difficulty can be met by supposing that all or most of the people of the man's two environments reported to him similar experiences of

---

[1] [2], p. 141. 'Your wife says, "You were very restless last night. What were you dreaming about?"'

double lives to those which he experienced. One way is to suppose that 'everyone's dream-life was coherent but that no person's dream-life corresponded with anyone else's' ([2], p. 143). This position formally meets the objection which I have made, but one might still wonder whether this description was a sufficiently coherent one. If everybody's lives were so different from those of others one might well wonder if there was sufficient coherence about the world for a public language applied to it to make sense. However, a coherent situation will certainly arise 'if we suppose that the dreams of everyone in England reveal a coherent order of events in our mystical lake district and let everyone have one and only one correlated lake-dweller whose waking experiences are his dreams' ([2], p. 143). If one supposes in this way that one meets in the new environment people of the old, and that the possibility of travel between the two environments is recognised in each, this objection would seem to have been met. Quinton describes such a situation as a possible but inessential variant on his myth, but the considerations which I have adduced would seem to suggest that it is a highly desirable, if not essential, variant, if the myth is to do the job for which he intends it. In Quinton's variant:

there are various ways in which we can suppose that people who know each other in England could come to recognise one another at the lakeside, for example, by drawing self-portraits from memory or by agreeing, when in England, to meet at some lakeside landmark. The injury which I do you at the lakeside may be revenged not there but in England ([2], pp. 141f).

Under these conditions anyone who appeared to find himself at different instants in the two different environments would have every reason to say that both environments were real.

But at this point a third objection naturally arises. This is the objection that sameness of body is necessary for sameness of person, a view powerfully advocated in recent years by Bernard Williams (especially in [5]). On this view P' at a later time t' is the same person as P at an earlier time t only if P' has the same body as P. Sameness of body however is a matter of the normal conditions for sameness of material objects, including the conditions of spatio-temporal continuity relative to a previously identifiable frame, described on pp. 19ff. For B' to be the same body as B, there has to be continuity between the two bodies. But the identity of bodies on Earth is assessed relative to E-frames. A body which moved to another space would not preserve spatio-temporal continuity relative to an *E*-frame; and so no body could move to another space. Hence, on this view that bodily continuity is a necessary condition of personal identity, neither can persons move to another space.

We do of course normally suppose that if *P'*'s body is not continuous with that of *P*, then *P'* is not the same person as *P*. But we do also have other kinds of evidence for personal identity. If *P'* has the same apparent memory as *P* (i.e. claims to remember as happening to him many and various things which in fact happened to *P*) and a similar character, then that is good grounds for supposing that *P'* is *P*. (This is in effect the satisfaction of our criterion of similarity, as described on p. 21.) But what are we to say if this criterion of similarity of memory and character is very well satisfied, but the criterion of bodily continuity is not satisfied? Williams answers that in that case the persons are not the same.

Suppose a man, whom Williams calls Charles, changes his character and claims to be Guy Fawkes. We investigate his claim and our inquiry turns out:

> in the most favourable possible way. . . . Not only do all Charles's memory claims that can be checked fit the pattern of Fawkes' life as known to historians, but others that cannot be checked are plausible, provide explanations of unexplained facts and so on. Are we to say that Charles is now Guy Fawkes, that Guy Fawkes has come to life again in Charles's body, or some such thing? ([5], pp. 237f.).

No, answers Williams, because:

> if it is logically possible that Charles should undergo the changes described, then it is logically possible that some other man should undergo the same changes, e.g. that both Charles and his brother Robert should be found in this condition. What should we say in that case? They cannot both be Guy Fawkes; if they were, Guy Fawkes would be in two places at once, which is absurd. We might say that one of them was identical with Guy Fawkes, and that the other was just like him; but this would be an utterly vacuous manoeuvre since there would be *ex hypothesi* no principle determining which description was to apply to which. So it would be best, if anything, to say that both had become like Guy Fawkes, clairvoyantly knew about him, or something like this ([5], pp. 238f.).

What this shows is that in the simpler case where only Charles claimed to be Guy Fawkes, it would be 'vacuous' ([5], p. 240) to say that Charles was Guy Fawkes, for there would be no difference between saying that Charles was Guy Fawkes and saying that he was exactly similar to Guy Fawkes.

The most general justification which Williams provides of the last stage of this argument is that 'identity is a one-one relation, and that no principle can be a criterion of identity for things of type T if it relies on what is logically a one-many or many-many relation between things of type T' ([7], pp. 44f). If we say in the simpler case (where only Charles, but not Robert, claimed to be Guy Fawkes) that Charles is Guy Fawkes, we are using a criterion which could lead us to contradict ourselves. Williams demands that a criterion of identity should be such that rigorous application of it should of logical necessity never lead to self-contradiction; in particular, consistent application of it should of logical necessity never lead us to say $A = B$ and $A = C$ but $C \neq B$. If we say that Charles is Guy Fawkes, we could on exactly similar grounds have to say that Robert is Guy Fawkes, but since Charles is not Robert, we would have contradicted ourselves. The criterion of bodily continuity, Williams suggests, does ensure that only one later person can be the same person as an earlier person, for only one person can have his body; whereas very many later persons could satisfy the memory and character criterion.

In the past twenty years there has been a very considerable philosophical literature on the subject of personal identity. Many writers sympathetic to Williams' general approach have modified it in one important respect. The continuity which guarantees personal identity is not, they claim, continuity of the whole body but of that part of the body which is causally responsible for person's memory and character — viz. continuity of brain. So if A's brain is taken out of A's body and put into B's body and B's brain is taken out of B's body and put into A's body, and the two brains are connected up with their new bodies so that we have fully functioning persons, then thereafter the person who had B's body would be A, and the person who had A's body would be B. There would be such continuity of memory and character that we would let brain continuity dictate who we said was A, rather than continuity of the rest of the body.

But a major difficulty with Williams' original view and this revised view is that in fact, as expanded so far, neither of them guarantee the uniqueness for which Williams sought. This is especially evident in connection with the revised view that brain-continuity is a necessary condition of personal identity. The human brain consists of two hemispheres. It is known that if one or other hemisphere is damaged in an accident, the person may continue to live and show much normal human behaviour. So it is logically possible (and much more likely to occur than the Guy Fawkes story related by Williams) that *A*'s brain could be taken out of *A*'s body, and the right hemisphere be placed in one empty skull, the left hemisphere be placed in another empty skull, and both half-brains be connected up so that we have two fully functioning persons. Let us call the new persons *B* and *C*. Both *B* and *C* have brain-continuity with *A* (and so, we may reasonably suppose, behave like *A* and make memory claims similar to those which *A* made). By the criterion of brain continuity, as so far expanded, both would be *A*. The duplication which Williams saw as damaging the memory and character criterion is equally damaging to the brain continuity criterion.

In the face of this difficulty exponents of the latter view have usually added a clause to the effect that if two later persons both satisfy the criterion of brain continuity for being the same person as a previous person, then neither are that person. For example, David Wiggins suggested that we

analyse *person* in such a way that coincidence under the concept *person* logically required the *continuance in one organized parcel of all that was causally sufficient and causally necessary to the continuance of essential and characteristic functioning, no autonomously sufficient part achieving autonomous and functionally separate existence*. ([9] p. 55. His italics)

That is, to be the same person, you have to have a bodily organ causally necessary and sufficient for memory and character, i.e. some of the same brain, and no part of that brain must animate another person.

However, as Wiggins appreciates, there is in a definition of this kind the following awkwardness, that a man's identity (i.e. whether or not he is identical with a certain past person) could, as a matter of logical necessity, depend on the success or failure of an operation to a brain other than his own. For suppose, as in the previous story, that *A*'s brain is taken out of *A*'s body and the right hemisphere placed in one empty skull and the left in another. Suppose that the transplant of the right hemisphere takes and that we have a fully functioning person, *B*. If the other transplant fails, there will be only one part of the original brain animating a person, and so on a Wiggins-type account that person will be identical with the original person. *B* will be *A*. But if the other transplant takes neither *B* nor anyone else will be *A*. But it seems logically absurd to suppose that who B is depends (as a matter of logic) on what happens as the result of an operation in a body other than his own, or that I can survive an operation only if some transplant fails.

Something has gone badly wrong in this kind of theory of personal identity, and two different kinds of theory of personal identity have been developed in recent years which do meet this kind of difficulty. One is a theory which takes the approach of Williams and Wiggins much further by stressing the similarity of personal identity to the identity of other kinds of material objects. For desks and ships, for armies and countries, our criteria of identity are not rigorous enough to give unique answers in all cases. We saw in Chapter 1 that there was no unique true answer as to which ship was the ship of Theseus. Sometimes the answer that B is the same object as an original object A is as near to the truth as an answer that a different object C is A. If we want uniqueness we may need arbitrary decisions, especially in cases where there is any extensive transplantation of parts — whether they be parts of ships or parts of persons. A number of writers have stressed this, but one writer in this tradition who has been much discussed is Derek Parfit [11]. He goes on to claim that what is important is not so much personal identity, but what he calls psychological continuity. Very roughly the latter is similarity of memory and character causally linked. Thus a person P' at time t' is psychologically continuous with P at t, if P' remembers what P remembers and P's memory causes (e.g. via continuity of brain matter) P'''s memory. Now two later persons cannot both be identical with P, but two later persons can both be psychologically continuous with P, and so P can 'survive' as both, although not be identical with both. Psychological continuity is a matter of degree, since similarity of memory (and character) and extent of causal dependence can vary from total to zero. Personal identity in general involves psychological continuity (or, perhaps in its absence, bodily continuity) but psychological continuity can hold where there is no identity. Two persons at different times are the same person 'if they are psychologically continuous and there is no person who is contemporary with either and psychologically continuous with the other' ([11] p. 13).

On a theory of personal identity such as that of Williams and Wiggins some sort of spatio-temporal continuity relative to a previously identifiable frame is necessary for personal identity, and hence movement of a person from one space to another is not possible. On Parfit's view the more important concept of psychological continuity involves causality. I do not see any need for him to hold that causality involves spatio-temporal continuity. But he might well hold this, and if he did again motion between spaces would not be logically possible (since the identity of persons involves psychological continuity, or perhaps in its absence, bodily continuity).

It seems to me however that Parfit's approach begs crucial factual questions and is very difficult to make coherent. In our transplant story look at the matter from A's point of view, before the operation. Suppose the person in B's body is to be tortured and the person in C's body is to be rewarded. A wishes to be rewarded and not to be tortured, but has he cause for hope or fear? That depends whether he will experience the feelings of the body into which his left brain hemisphere is transplanted, or the feelings of the body into which his right brain hemisphere is transplanted, or of neither, or of both. Suppose that both transplants take. In that case Parfit seems to suggest that the last suggestion is right; that A will at any rate in some measure have the feelings both of B and of C. But it is hard, to put it mildly, to make sense of how that can be, since B and C will not experience each other's feelings. Each of the other three suggestions seems coherent enough, seems to contain no self-contradiction. In the first case, although both B and C would behave like A and make somewhat similar memory claims to A, only C would be A. A's hopes to be rewarded would be fulfilled. In the second case, although both B and C would behave like A and make somewhat similar memory claims to A, only B would be A. A's fears about being tortured would be fulfilled. In the third case despite the similarity of memory and character, neither B nor C would be A. Each of these seems to be a factual possibility. Yet how can logic tell us which will happen? In the circumstances described, surely one and only one suggestion would provide the true description of the situation, and yet we would not know which was the true description, nor might we have any idea about how to go about finding out which was the true description.

All of this suggests a very different theory of personal identity which along with others I have myself advocated in recent years (e.g. in [12]). On this view personal identity is something ultimate, not further analysable. It is not *constituted* either by similarity of memory and character or by bodily continuity, although it is a basic inductive principle (analogous to the much wider criterion of similarity discussed in the last chapter) that similarity of memory and character and continuity of body are strong evidence that two persons are the same and the lack of such similarities and continuity is strong evidence that two persons are not the same. If P' at time t' does not have a body continuous with that of P at t, that is good grounds for supposing that P' is not the same person as P. It is not however conclusive grounds. They could be the same person, and the very good satisfaction of the other criterion (similarity of memory and character) would be good grounds for supposing that they are the same, especially if there was no rival candidate P'' at the same time as P' who satisfied the criterion of similarity of memory and character equally well. If the criterion is merely evidence of identity, and not constitutive of it, no logical difficulties arise. The logical possibility of a Robert turning up does not make the similarity of Charles to Guy Fawkes any less evidence of his being Guy Fawkes. If Robert does in fact turn up, we do not know which of Charles and Robert is Guy Fawkes. They cannot both be, but one of them may be — despite our ignorance.

This theory of personal identity has the consequence that the survival conditions for persons are very different from those for other material objects. This is not surprising because cars and ships do not have feelings or hopes to survive, and therefore there is nothing puzzling about there being two later ships, each of which is as much the same ship as the earlier ship as is the other ship. But this does seem an incoherent suggestion when we come to talk about persons. On this theory of personal identity, continuity of body and so spatio-temporal continuity relative to a previously identifiable frame is not a necessary condition of personal identity, and so persons may move between spaces. Evidence of identity of persons in a new space with persons in an old space would be provided by very strong satisfaction of the criterion of similarity and memory by one and only one person P' in the new space for each person P in the old space.

Finally, I consider a fourth objection which may be made to Quinton's myth. It claims that as Quinton describes the situation, we do not have adequate evidence for believing that the two 'places' belong to two different spaces, even though they may in fact do so. It might be granted that both 'places' were real places, and that men passed from the one to the other discontinuously. But an objector could ask, what real reason have we for supposing that the second place is not spatially related to the first? Quinton's answer here is brief. 'Suppose that I am in a position to institute the most thorough geographical investigations and however protractedly and carefully these are pursued they fail to reveal anywhere on earth like my lake. But could we not then say that it must be on some other planet? We could, but it would be gratuitous to do so. There could well be no positive reason whatever, beyond our fondness for the Kantian thesis, for saying that the lake is located somewhere in ordinary physical space, and there are in the circumstances envisaged, good reasons for denying its location there' ([2],

p. 143). Quinton's 'good reasons' are presumably that we have in the one environment looked for the place of our other-life and not found it. But having looked for a thing and not found it is fairly weak reason for believing it not to exist unless we have looked at most of the places where it could possibly be. An objector might well urge in such a case as Quinton has described that the place of the other life is just as likely to be beyond the furthest observed galaxy as to be not spatially related to ourselves, and hence that it is not reasonable to postulate the odder alternative. There could be two kinds of further good reason which might make Quinton's alternative the more reasonable one to adopt.

The first kind of good reason would be that the laws of nature observed to hold in one life were markedly different from those observed to hold in the other. Thus, suppose the men transported from one environment to the other include scientists. They observe that in one life an inverse square law of gravitational attraction holds, while in the other life an inverse cube law holds, without there being any obvious environmental differences (e.g. vast density of matter in one life) to which the difference of law could be attributed. This difference holds in all of many observed regions in each life. The scientists are now faced with two alternatives. If they claim that the two environments are both in the same space, they have to admit a discontinuity of scientific laws, that one law holds in one environment and in the other environment a different law holds; at the boundary between the two regions matter becomes subject to a discontinuous change of behaviour (there being no evidence within each region of any gradual change of law). To adopt this hypothesis would thus be to introduce an awkward complexity into science, and the scientist following the principle of scientific methodology that he ought to adopt the simplest hypothesis consistent with observations might well prefer the (admittedly provisional) hypothesis that the two places had no spatial relations to each other. Spatial discontinuity of law operation is something that the scientist has never since the seventeenth century allowed into science (on Aristotelian physics there was just such a discontinuity: one set of laws governed terrestrial phenomena, and another set celestial phenomena).

The second kind of good reason for adopting Quinton's alternative would, if it applied, be much stronger than the first. It might turn out that the space of at any rate one of the lives was a closed or finite space;[1] that is, contained only a finite number of places of finite volume, so that

[1] For more adequate characterisation of a closed space and of the grounds for claiming that a space was a closed space, see Chapter 6.

in whatever direction one set off from a given place, one would eventually arrive back at where one started. Now in that case one could in principle fully explore all the places in the space of one of the lives and failure to find among them the places of the other life would be very good evidence indeed for the former not being spatially related to the latter. Even if the space was too big to be fully explored in practice, we could explore some of it and our evidence of failure to find the other environment would be much better evidence of its non-existence in that space if the space were a finite one than if the space were an infinite one. Further, although we might not be able fully to explore all the regions of the finite space, we might be able to learn quite a lot about them by telescopes and similar devices so as to have quite reasonable evidence that the other environment was not in that space.

I conclude that we could have good evidence that the two environments were not spatially related and hence that they belonged to two different spaces.

Thus the four possible objections to Quinton's original thesis can be rebutted and the thesis stands. It is not a necessary truth that there is only one space; and that there is more than one space is something of which men could have evidence.

In order for a man to pass from one space to another, he must have a means of travel other than the sort of way by which we normally get to places, which, to give it a name, I propose to call local motion. In local motion between two places A and B I can travel by many routes, one of them being a shortest line. If I travel by local motion on a shortest line from A to B, then as I move I get further away from A and nearer to B, in the sense that a man would have to lay more and more measuring rods to cover the distance along the line between myself and A and fewer and fewer to cover the distance along the line between myself and B. All of this holds if I travel on foot, by bus or by rocket or — *mutatis mutandis* — if it is not I that is travelling but a marble or a bullet. If travel by local motion is along any other line than a shortest line then distance from any place on the line to A or B can be measured, and in my travel from A after I have reached a certain place on the line I get nearer and nearer to B.

To get to another space we would have to use a means other than local motion, one lacking the characteristics described above. Quinton's method of getting there by falling asleep clearly satisfies this demand, but there are many other methods which would do so. You might only need to recite the opening words of the Koran and in a flash your surroundings are entirely new. To get back again you recite the closing words and in a flash you are in the old surroundings. Here again there is no question of travelling along a line, distance from places along which to the starting or finishing points could be measured.

Other spaces are not merely a matter of philosophical thought-experiment. The traditional Christian doctrine of Heaven may most plausibly be regarded as a doctrine about another space. Christians have always maintained that in Heaven after the General Resurrection men will be embodied (though these bodies will lack many of their earthly limitations). Bodies must be located at places. So Heaven must be a place. The medievals in general believed that the Heaven of the blessed was situated just beyond the sphere of the fixed stars. But seventeenth-century astronomy showed that view to be mistaken. There is no 'sphere' of fixed stars. The telescopes that pry daily ever deeper into the Universe have revealed nothing resembling Heaven. Even if the Christian maintains that Heaven lies far beyond the furthest observed galaxy, there is still the difficulty that different laws of nature are said to hold in Heaven, different from those which hold within the observable Universe (bodies are freed from their 'earthly limitations'), and this, as we have noted, makes awkward the assertion that both belong to the same space. If the Christian wishes to maintain the doctrine that Heaven is a place, he does much better to claim that it is a place not spatially related to Earth. Such a doctrine is, we have urged in this chapter, a logically possible one; and indeed, although this is not my thesis here, one much more consistent than the medieval view with much traditional Christian theology.

The claim of this chapter has been that it is not a logically necessary truth that there is only one space. However, unless we accept the Christian or some other such metaphysic, we have no reliable information about any space except our own, and the remainder of the book will be concerned with the actual and possible properties of this space.

BIBLIOGRAPHY

[1] Kant, *Critique of Pure Reason* (trans. N. Kemp-Smith), London, 1929, Transcendental Aesthetic, section 1, 'Space'. (In references to this work in this and other chapters, where a passage appears in both editions I give a reference to the second edition only. A followed by a number gives the page in the original first edition; B followed by a number gives the page in the original second edition.)
[2] Anthony Quinton, 'Spaces and Times', *Philosophy*, 1962, 37, 130–46.

For definitions of discreteness, density, and continuity see:

[3] E. V. Huntington, *The Continuum and Other Types of Serial Order*, Second Edition, Cambridge, Mass., 1917.

For discussion of the point that empirical tests could not decide between the theory that space is continuous and the theory that it is merely dense, see:

[4] W. Newton-Smith, 'The Under-Determination of Theory by Data', *Proceedings of the Aristotelian Society*, Supplementary Volume, 1978, 52, 71–91.

For rival theories of personal identity see:

[5] Bernard Williams, 'Personal Identity and Individuation', *Proceedings of the Aristotelian Society*, 1956–7, 57, 229–52.

[6] Robert C. Coburn, 'Bodily Continuity and Personal Identity', *Analysis*, 1960, 20, 117–20.

[7] Bernard Williams, 'Bodily Continuity and Personal Identity', *Analysis*, 1960, 21, 43–8.

[8] Anthony Quinton, 'The Soul', *Journal of Philosophy*, 1962, 59, 393–409.

[9] David Wiggins, *Identity and Spatio-Temporal Continuity*, Oxford, 1967.

[10] Bernard Williams, 'The Self and the Future', *Philosophical Review*, 1970, 79, 161–80.

[11] Derek Parfit, 'Personal Identity', *Philosophical Review*, 1971, 80, 3–27.

[12] Richard Swinburne, 'Personal Identity', *Proceedings of the Aristotelian Society*, 1973–4, 74, 231–47.

Some of the above articles are republished, together with extracts from the writings of classical philosophers in:

[13] John Perry (ed.), *Personal Identity*, Berkeley and London, 1975.

3 Absolute Space

Places, we have seen, are identified by their spatial relations to material objects forming a frame of reference, and whether spatial things are said to be at rest or in motion and how far they are said to have moved depends on the frame chosen.

The question arises whether in some sense one frame of reference is a more basic frame than any other, so that motion relative to it may be described as absolute motion, and motion relative to all other frames be described as mere relative motion? Many philosophers and scientists have believed that there is such a frame and they have termed location by reference to it location by reference to Absolute Space, and motion relative to it motion through Absolute Space.

A classical advocate of Absolute Space was Newton. He wrote:

Absolute Space, in its own nature, without relation to anything external, remains always . . .[1] immovable. Relative Space is some movable dimension or measure of the absolute spaces; which our senses determine by its position to bodies; and which is commonly taken for immovable space; such as is the dimension of a subterraneous, an aerial, or a celestial space, determined by its position in respect of the earth [3].

Absolute Space is thus Space, the places constituting which are reidentified by the frame of reference which does not really move. Relative Space is Space, the places constituting which are reidentified by a frame of reference which does really move. We must now investigate the propriety of the concept of Absolute Space.

Clearly some frames are more basic frames than others. For relative to the Earth, the Sun moves round the Earth, and, relative to the Sun, the Earth moves round the Sun; but we say that really it is the Earth that is moving and not the Sun. Why do we say this? If we take the Earth as

[1] I have omitted from this place the words 'similar and'. I ignore them for the present and discuss them on p. 57.

fixed and try to set down the laws of mechanics, and especially the mechanics of planetary movement, in terms of motions relative to it, they turn out to be very complicated laws. The vast complexity of astronomy before Copernicus arose partly from taking the Earth as fixed. Yet if we take the Sun as fixed (in the sense that it continues to occupy the same place; we allow it to rotate), the movements of the planets and of so many celestial bodies can be explained very simply in terms of Kepler's three laws of planetary motion; and these — taking the Sun as approximately fixed — can be explained very simply in terms of Newton's three laws of motion and his law of gravitational attraction, laws which also explain a host of other phenomena. To say that one frame is more basic than another frame is thus to say that the simplest laws which can be formed explaining the observed behaviour of bodies relative to it are simpler than the simplest laws which can be formed explaining the observed behaviour of bodies relative to any other frame. To say that one proposed set of laws L is simpler than another set L' is to say that on balance L uses a simpler mathematics or other symbolism than L', contains fewer mathematical terms, postulates fewer and less mysterious unobservable entities, and forms a more coherent system, so that odd coincidences and exceptions allowed by L' find a neat and natural explanation by L, whereas the converse does not so much occur.

Compatible with any finite set of phenomena there will always be an infinite number of possible laws, differing in respect of the predictions they make about unobserved phenomena. Between some of these ready experimental tests can be made, but experimental test between others is less easy and between them we provisionally choose the simplest hypothesis. Evidence that a certain suggested law is simpler than any other is not merely evidence that it is more convenient to hold that suggested law than any other, but evidence that the suggested law is true. Thus, to take a trivial example, if I in Yorkshire in 1968, have observed many swans and noted that they are all white, this is evidence in favour of the truth of the hypothesis that all swans are white rather than, for example, the hypothesis that all swans are black except those observed in Yorkshire in leap years which are white. The latter is clearly a more complicated hypothesis than the former, being less coherent, and since both give correct predictions within the tested range, the truth of the former is better substantiated than the truth of the latter.

Or, to take a less trivial example, if up to 1984 I have observed various phenomena predicted by Einstein's General Theory, this is evidence in favour of his theory rather than a mathematically much more complicated theory which also predicted those phenomena, but which predicted different phenomena to be observed thereafter from those predicted by General Theory. If we supposed that a more complicated hypothesis was just as likely to be true as a simple one, if both are compatible with phenomena observed so far, then we should have to say that we had no grounds whatever for claiming that any hypothesis compatible with data observed so far was more likely to be true than any other and so no grounds whatever for making predictions about the future.

A number of writers, following Reichenbach (*Experience and Prediction*, Chicago, 1938, § 42), have distinguished between 'descriptive simplicity' and 'inductive simplicity', claiming that if a hypothesis h_1 has greater descriptive simplicity than h_2, that does not make it more likely to be true, whereas if it has greater inductive simplicity it does make it more likely to be true. In the former case h_1 is just a simpler formulation of the same claim about the world as h_2. Whereas in the latter case, although h_1 and h_2 may be equally compatible with observations made so far, they make different claims about the world outside the region observed, and the greater simplicity of h_1 has the consequence that the claims of h_1 about the latter are more likely to be true than those of h_2. Now quite clearly if h_1 really is just another formulation of h_2, each is as likely to be true as the other. But what is at stake is when is h_2 just a reformulation of h_1, and when is it making a different claim. Reichenbach and his followers agree that h_1 and h_2 are making different claims, i.e. are different statements, when they make different predictions, the occurrence or non-occurrence of which an observer can observe straight-off, as in the example about the swans in Yorkshire. Yet is it possible that h_1 and h_2 should make identical testable predictions, yet differ in respect of some other untestable claims about the world, and so be different theories? For a positivist the answer must be No; for, as we saw in the Introduction, he equates the factual with the verifiable. But, we also saw, there is no reason to affirm this equation. There may be claims making identical testable predictions yet differing in their factual content. An example would be where h_1 is the present theory of fundamental particles (protons, electrons, etc.) and h_2 is the theory which makes exactly the same testable predictions as h_1 and yet denies that there are any protons, electrons, etc. whose movements explain what we observe. The grounds for judging h_1 more likely to be true than h_2 are it has a few simple laws of particle motion of which the many and complex lower-level laws about observables are consequences, whereas h_2 has only the latter. The situation is similar for different laws of mechanics L and L' related to different frames of reference. Their observable consequences may be the same but they may differ in the forces which they postulate to explain those consequences. The fact that one postulates much simpler forces than the other is ground for supposing that those are the forces which operate in nature. Thus Newtonian mechanics distinguished between forces which are really in nature (e.g. the gravitational force) and those which merely seem to operate if you take an accelerating frame of reference (e.g. the centrifugal force, or the Coriolis force).

Suppose now that one set of laws L from which explanations of observed phenomena of some type can be derived is simpler than another set L' from which they can also be derived, and L uses a frame of reference F for reidentifying places, while L' uses a frame F'. Suppose too that L is the simplest set of laws which can be formulated for the phenomena using F, and L' the simplest set using F'. Then since a law being simpler than another is a reason for believing it more likely to be true, we have reason to believe that L is more likely to be true than L'. Hence we have reason to believe that F gives a more proper way of reidentifying places than F', that F is a more basic frame of reference than F'. Hence when comparing two frames, one of which is more basic than another, we say that it is the less basic one which really moves and is responsible for the relative change of position, whereas the motion of the other is 'merely relative'. In order to ascertain the laws of nature we need a clock as well as a spatial frame of reference. For each frame of reference being compared we choose provisionally the clock which yields the simplest laws of nature relative to that frame — in a way and subject to a condition (my first criterion on p. 178) to be described fully in Chapter 11. But whatever clock we use, the path traced out by the planets relative to the Earth is much more complicated than that traced out relative to the Sun, and hence the Sun forms a more basic frame than the Earth.

So the question of whether there exists Absolute Space reduces to the question whether there is a frame of reference more basic than any other. And by 'a frame' we should, I suggest, mean 'any actual or physically possible frame'. If it were found that of all frames of reference formed by existing material objects one was more basic than any other, but there were reasons to believe that if another frame were constructed moving in a certain way relative to the former it would be as basic as the previous one, we should hardly want to say that there was Absolute Space. For Absolute Space is not the sort of thing that can exist today, be gone tomorrow, and return again the next day — according to whether or not we send a satellite into space. It cannot be a mere technological accident that there is Absolute Space. So to say that there is Absolute Space is to say that some actual or possible frame is always more basic than any other. Actual frames might become or cease to be most basic frames, and different actual frames at different temporal instants might be stationary in Absolute Space. For clearly any frame consisting of material bodies could disintegrate, while Absolute Space presumably could not; and any frame could be moved relative to anything else, including, presumably, Absolute Space. Even if today the Earth were stationary in Absolute Space, the Earth could be blown up or set in motion tomorrow and then we would need a different frame to identify Absolute Space. But to claim that there is Absolute Space is to claim that a frame can be postulated (one which may or may not at any temporal instant or at all temporal instants correspond to an actual frame), and its varied motions relative to actual bodies at various times described such that it remains always the most basic frame of reference.

The question of the existence of Absolute Space thus becomes a question for scientific theory. Up to the sixteenth century most men considered that there was Absolute Space and that it was determined at all instants of time by the one basic frame of reference which was the Earth. They repeated Aristotle's view that 'the Earth does not move nor does it lie anywhere except in the middle' of the Universe ([1], 296 b). The arguments which Aristotle [1] presented in favour of this view were basically arguments of simplicity, although he would not have described them in this way. Thus among the arguments from astronomy and mechanics which Aristotle gives in *On the Heavens* for the absolute immobility of the Earth is this one from astronomy.[1] The planets are observed to retrogress at regular intervals, viz. to travel temporarily in a direction opposite to their normal direction relative to the 'fixed stars'. The explanation of this which Aristotle accepted, because it was the simplest one known to him, was that the planets are subject to two circular motions, one progressive (from west to east) and the other retrogressive (from east to west), the combination of which normally produces progression but sometimes retrogression. If we are to suppose that the Earth revolves around something and so behaves like a planet, we must suppose that it also has these two motions. Again, though Aristotle does not bring this out, the only reason for making this supposition is that it is simpler to suppose that all the planets have the same kind of motion than that one has a different kind of motion from the others. But if the Earth

---

[1] [1] 296a–b. For interpretation see the note on p. 242 of the edition in The Loeb Classical Library by W. K. C. Guthrie, London, 1960.

did have two such motions, the 'fixed stars' would appear sometimes to retrogress, for, when the Earth retrogressed, to an observer on it the 'fixed stars' would appear to retrogress. 'Yet this is not observed to take place.' So, Aristotle is arguing, our astronomical theory becomes simpler if we suppose that the Earth does not move from the centre of the Universe. A similar difficulty arises if we suppose that the Earth, although remaining at the centre of the Universe, rotates on its axis. Here too, for simplicity's sake, we ought to suppose, he urges, that the rotation is in the plane of the ecliptic; but this would produce irregular motions of the fixed stars, and these are not observed.

When Copernicus, Kepler, and Galileo argued in the sixteenth and seventeenth centuries that the Earth revolved annually round the Sun and rotated daily on its axis, they were more explicit in putting forward the increased simplicity of their scientific theory as the grounds for holding it to be true. Copernicus claimed that he could provide an explanation of how the planets moved on spheres having uniform circular motion 'with fewer and much simpler constructions than were formerly used',[1] if some assumptions, including the annual revolution of the Earth around the Sun, were granted him.

Copernicus was hardly justified in making this claim for his astronomical system was very clumsy, but Kepler undoubtedly was justified. The thirty to forty spheres of Ptolemaic astronomy with their inaccurate predictions were replaced by Kepler's three neat and accurate laws of planetary motion. Hence it was simpler to take the Sun rather than the Earth as fixed.

The defenders of the old view argued that although the astronomy might be simpler, terrestrial mechanics became mysterious. Galileo and later followers countered their objections by showing that a new mechanics could be constructed which was simpler than the old one. His opponents, for example, argued that if the Earth was rotating, and you dropped a ball from a tower then it should fall a large number of yards to the west of the base of the tower, since the Earth and so the tower on it would have passed to the east that much during the time of fall. The fact that this did not happen indicated that the Earth was absolutely stationary. Yet according to Galileo's mechanics the ball when released would have as well as a downward velocity the velocity of the tower from which it was released, and hence — approximately — it should arrive in the same place at the bottom of the tower, whether the Earth was

---

[1] Nicholas Copernicus, *Commentariolus* in *Three Copernican Treaties* (trans. and ed. Edward Rosen), Dover edition (New York, 1959), p. 57.

moving or not. Galileo backed up this view by urging that a stone
dropped from the top of the mast of a moving ship has as well as its
downward motion, the ship's motion, which is why — approximately —
it falls to the bottom of the mast.[1] A simple explanation of the latter
phenomenon followed from Galileo's mechanics, whereas you would
have to complicate the old system to account for it. The principle of
Galileo's mechanics used above was generalised by Huygens and
Newton as the principle of mechanical relativity — that the same laws of
mechanics hold in any frame of reference moved with uniform
rectilinear motion relative to a given frame. A frame of reference relative
to which Newton's laws of motion hold is known as an inertial frame.
This is — to a high degree of approximation — a frame stationary or in
uniform motion relative to the 'fixed stars'[2] or — less approximately —
relative to the Sun, or — to a yet smaller degree of approximation — for
purposes of small-scale mechanical experiments, the Earth (since its
motion is during periods of time a few seconds long approximately
rectilinear). Newton's laws of motion were, it was held, the simplest
which apply to any frames. Consequently if we consider only mechanics
there will be an infinite number of frames relative to which scientific laws
are highly simple, and none relative to which they are yet more simple.
With whatever uniform velocity the ship moves relative to the Earth, the
same laws of mechanics will hold on it. Of two bodies moving uniformly
relatively to each other, mechanics cannot decide which is 'really
moving'.

   This showed that the simple conclusions of Copernicus, Kepler, and
Galileo himself were unjustified. One could not conclude that the Sun
was absolutely still, only that it was either absolutely still or in absolute
uniform motion (viz. in uniform motion relative to Absolute Space). Yet
for two bodies moving non-uniformly relative to each other, Huygens
and Newton showed that the laws of mechanics would differ, dependent
on which one took as the frame of reference, and the laws of mechanics
relative to an inertial frame would be simpler than laws relative to a non-
inertial frame (viz. one with non-uniform motion relative to an inertial

---

[1] For these arguments see [2], pp. 142–5.
[2] Since the stars preserve over short periods of time very similar spatial relations
to each other they can form a rough frame of reference. It was realised in the
eighteenth century that the stars are in fact in relative motion, and so more
complicated techniques are needed to identify to any very high degree of
approximation the class of inertial frames. For the various methods used today
see G. M. Clemence, 'Inertial Frames of Reference', *Quarterly Journal of the
Royal Astronomical Society*, 1966, 7, 10–21.

frame). An important case of non-uniform relative motion is revolution around a central body. As we have seen, the laws of planetary motion and of the mechanics which explains that motion become simpler if we take the Sun as the frame of reference rather than the Earth. This is because the Sun is to a far greater degree of approximation than is the Earth an inertial frame. Mechanics sometimes enables us to say of two frames in relative acceleration that one is more basic than another. So it enables us to say that the Earth and not the Sun is really moving. But it never enables us to say, Huygens and Newton showed, of two frames in uniform relative motion that one is the more basic.

Since the inertial frames where the most basic frames detected, Newton supposed that one of them must be stationary relative to Absolute Space. Hence any body accelerating relative to an inertial frame would, he reasoned, be accelerating relative to Absolute Space and hence he termed the acceleration absolute acceleration. Absolute acceleration was detectable, but absolute velocity was not.[1]

It was this fact that absolute acceleration, understood as acceleration relative to an inertial frame, could be detected, that led Newton to suppose that Absolute Space existed. Newton's disciple Clarke conducted a celebrated controversy with Leibniz about Absolute Space. Leibniz argued, as I have done, that Newton's science showed that there was no Absolute Space, because were 'The Universe moving forward in an infinite empty space', 'then would happen no change, which could be observed by any person whatsoever.'[2] In rebuttal Clarke argued 'that a

---

[1] Newton has an argument in *Principia*, Book iii, Hypothesis i, and Proposition xi, Theorem xi, to show that the common centre of gravity of the Earth, Sun and planets is at rest. The argument seems to be that, since that centre and the 'fixed stars' retain their distances apart, then if the common centre of Earth, Sun, and planets moved, the centre of gravity of the Universe would move. But the latter is impossible, he claims, and its impossibility acknowledged by all. But why should it be impossible? On Newton's view of Space there can be no possible objection to the doctrine that the centre of the Universe is in uniform absolute motion. This he admits elsewhere in his writings.

[2] [4], pp. 63f. A modern philosopher would have gone on from that point to argue that in that case to claim that the Universe was moving rather than still was to make an unjustified assertion. Leibniz instead reasoned that since there would be no observable difference between the two states, the Universe at rest relative to Absolute Space and the Universe in motion relative to Absolute Space, 'such an action would be without any design in it', viz. God could have had no reason for bringing about one state rather than the other. There would be no 'sufficient reason' for the existence of one state rather than the other. Since all occurrent states needed a sufficient reason for their occurrence, neither of these two states could occur. Hence there could not be Absolute Space.

sudden increase or stoppage of the motion as a whole, would give a sensible shock to all the parts' ([4], p. 101). Leibniz was claiming that there was a large class of equally basic frames; Clarke that some frames were more basic than others. Both these claims were right. But Leibniz was surely correct to urge against Clarke and Newton that since it was impossible to detect whether or not a body was still or in uniform motion relative to Absolute Space, affirmation of the existence of Absolute Space was not justified. You could not on Newton's science distinguish between 'a sudden increase' and a sudden 'stoppage of the motion as a whole'.

Talk of Absolute Space still continued, however, while men cherished the hope that scientific laws other than those of mechanics would serve to reveal the most basic frame of reference. The laws of the propagation of light were the obvious candidate. But at the beginning of this century the experimental work leading up to and developed from the Special Theory of Relativity suggested (in a way to be described in Chapter 11) that light has the same velocity relative to all inertial frames. Hence the Principle of Mechanical Relativity was generalised as The Principle of Relativity to assert that all the laws of physics are the same in all frames in uniform motion relative to a given frame; but it was claimed, as before, that they take their simplest form in inertial frames. The Special Theory of Relativity (based on the principle of the constant velocity of light in all inertial frames) completed Newtonian mechanics in showing that there is no one privileged frame of reference, no frame more basic than all other frames, and so that there is no Absolute Space.

During this century the Special Theory of Relativity which purports to hold in a pure form only for empty Space has been seen by many as a special case of the General Theory of Relativity or one of a number of more developed cosmological forms of the latter or rival theories to it,[1] theories which purport to set forward the laws of nature for Space occupied by matter and radiation in a more adequate way than does either Newtonian mechanics or Special Relativity. It was Einstein's programme in setting forward General Relativity that 'the general laws of nature are to be expressed by equations which hold good for all systems of co-ordinates' ([8], p. 117), that is, that they should be 'generally covariant'. ('All systems' means, in terms of our Chapter 1, all E-frames.) However, almost all cosmological theories developed from or independently of the General Theory of Relativity postulate (on grounds to be described in Chapter 11) that matter is distributed on the

---

[1] See Chapter 14 for a fuller description of these.

large scale homogeneously throughout the Universe and hence that
there are a set of basic frames of reference relative to which the laws of
nature are simpler than relative to any other frames. These basic frames
are the 'fundamental particles' of the Universe, that is, the clusters of
galaxies which are the basic units of cosmology (or, more precisely — see
pp. 189f — those frames in the vicinity of each cluster having at most small,
temporary, and random velocity of recession relative to the cluster and
rotating relative to it in such a way that all other fundamental particles
at any given distance in any direction recede from them at the same rate).
The Universe is expanding, and so the fundamental particles are in
motion relative to each other — on some theories this motion is uniform
(motion being measured by change of 'proper distance' over 'cosmic
time')[1] and on others non-uniform. You can take any of the fundamental particles as fixed, and the laws of cosmology have the same form
relative to each of them. On the other hand, on almost all modern
theories, the laws of cosmology referred to any other frame of reference
would be markedly less simple. Hence there is on most cosmological
theories a class of frames equally basic with each other yet more basic
than any other frames. Such frames I will term in future equibasic
frames. But on any well established cosmological theory, as on the
Special Theory of Relativity, there is not a most basic frame, one frame
more basic than all others. There might have been. It is an empirical
truth that there is not. For, we have argued, talk of Absolute Space is
talk of bodies being absolutely at rest or in motion, and such talk is is
talk about one frame of reference being more basic than all others.

However, while on the cosmic scale there is no most basic frame of
reference, cosmological theory normally claims — on grounds to be
stated in Chapter 11 — that for regions of Space in the vicinity of any
galactic cluster, the fundamental particle (that is the frame which has the
motion described above relative to the cluster) is the most basic frame.
Cosmological theory supposes that laws of the behaviour of a local
galaxy (e.g. describing its rotation and dissolution) are simpler when
referred to this frame than when referred to any other. (Relative to this
frame, clusters consisting of a large number of galaxies will not normally
be in rotation. Clusters consisting of a solitary galaxy will, however,
normally rotate relative to the basic frame.) For instance, as will be
explained more fully in Chapter 11, relative to any of the other
fundamental particles, light in the vicinity of some particle will have a
velocity which varies in various ways. But relative to the particle in the

---

[1] See Chapters 5 and 11 for an exposition of the meaning of these expressions.

neighbourhood of which it lies, light sent in any direction will have a constant two-way velocity (i.e. average there-and-back velocity — approx. 300,000 km/sec). Yet although each fundamental particle forms a most basic frame of reference for the explanation of phenomena in its vicinity, no one particle will form a more basic frame of reference than any other. For although some phenomena can be explained more simply when referred to one particle than to others, other phenomena can be explained more simply when referred to another particle. Hence for the explanation of all phenomena no one particle forms a more basic frame of reference than any other.

In this discussion I have adopted the point of view that talk about Absolute Space is significant and have sought to elucidate what it means and what sort of evidence would show that there was such a space. There is however a tradition in philosophy which denies the significance or utility of talk about Absolute Space, and of this tradition Berkeley [5] and Mach [6] are the best-known exponents. Thus Mach, echoing earlier remarks of Berkeley ([5], p. 66) writes: 'No one is competent to predicate things about absolute space and absolute motion; they are pure things of thought, pure mental constructs, that cannot be produced in experience' ([6], p. 280). Now of course we cannot produce Absolute Space 'in experience' in the sense that we can point it out. It is not that sort of thing, nor are electrons or prime numbers; but we are perfectly 'competent to predicate things' about them. The fact that we cannot point something out does not show that it does not exist, nor that we cannot justifiably predicate things of it.

Mach does however have a more detailed argument designed to show that no evidence could confirm the existence of Absolute Space and hence, on his verificationist view, that the supposition of its existence was meaningless. He attempts to make this point by attempting to show that no experiment could demonstrate that a body was or was not rotating relative to Absolute Space. A rotation is an accelerated motion. Newton claimed, we have seen, that, while it was impossible to prove that a body was or was not in uniform motion relative to Absolute Space, it was possible to show that it was or was not accelerating relative to Absolute Space. Mach claimed, by considering Newton's celebrated bucket experiment, that Newton could not even make that distinction. Newton [3] had cited the following experiment. A bucket is filled with water and then rotated (relative to the inertial frame of the 'fixed stars'). At first, before the water has picked up the motion of the bucket, the water is flat. When it picks up the motion of the bucket, the surface of the water becomes concave. So we have two stages to the experiment. In the first stage the water and the bucket are in relative rotation, in the second stage they are not. So, Newton argued, mere relative rotation of water and bucket does not produce deformation of the water — for we have this relative rotation in the first stage without such deformation. There is relative rotation in the first stage, but only deformation in the second stage. Since deformation requires a force, and only absolute rotation would produce such a force, the water was rotating absolutely only in the second stage. Now Newton's argument which I have summarised is inadequate, since he does not justify his claim that deformation requires a force. In the second stage, as Mach pointed out, the water is still in relative rotation, but this time relative to the 'fixed stars'. We want to say that the water, and not the stars, is really rotating at the second stage, because mechanics would become excessively complicated if we supposed that the 'fixed stars' were caused to rotate absolutely by my twisting my hand. Hence the frame of the fixed stars is still or in uniform motion relative to Absolute Space, and the water in the second stage rotating absolutely as well as relatively. From the deformation during the period of absolute rotation, we conclude that such rotation produces a deforming force. This should be a conclusion derived from the view previously established that it is the water and not the stars which are rotating absolutely, not a premiss to prove that view.

Now having tidied up Newton's argument, let us pass to Mach's main criticism. Mach's criticism turns on the point that if we say that there is Absolute Space, we admit, as we saw on pp. 45 f, the possibility that a body, or all bodies, now stationary relative to Absolute Space, might come to move relative to it. We may have identified Absolute Space by a certain frame formed by actual material objects, but we must allow the possibility that that frame might come to move relative to it, and so it is to the space and not to any actual identifying frame that the laws of nature must be referred. Now if we say that the water at the second stage is rotating absolutely, then the law of nature must be formulated that the deformation of the water is caused by rotation relative to Absolute Space. This implies that if the stars were to rotate relative to Absolute Space and the bucket with the water in it remain still, then there would be no deformation. But how can we test this? We cannot rotate the stars but we could make the bucket thicker and see whether the mass of the rotating bucket produced a deformation of the water which remained still relative to the inertial frame of the fixed stars. If there were deformation, that would indicate that a rotation of the stars in Absolute Space about the Earth — recognisable by the stars ceasing to form an inertial frame — (if, *per impossible*, this could be achieved) would produce a similar deformation. But, Mach urges, 'no one is competent to say how the experiment would turn out if the sides of the vessel increased in thickness and mass till they were ultimately several leagues thick' ([6], 284). Mach is urging that the doctrine of absolute rotation commits us to the view that in such an experiment the mere rotation of the thickened bucket would not produce deformation, and whether or not such a result would occur we are not competent to say. The General Theory of Relativity and similar modern theories, unlike Newtonian mechanics and Special Relativity, assume that the mere rotation of the thickened bucket would produce deformation, but no satisfactory experiments have yet been done to show whether or not this assumption is justified.

But there is a conclusive objection to Mach's view here, even if we allow his verificationist assumption that to be factually meaningful a statement must be confirmable by observation. If the doctrine of absolute rotation does lead to definite testable, but so far untested, predictions, that is to credit it as a scientific theory. If Mach is contending that scientific theories should not extrapolate far from observable data, then that only shows an excessive positivist bias. The arguments against this position have been aired in many modern works on the philosophy of science, and I shall not repeat them here.[1] Of course if we have reason to doubt that its predictions would succeed, we

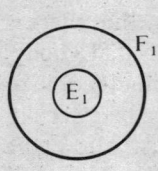

should not adopt the theory; and if we have reason to believe that they will succeed we should adopt the theory. If we have no reason either way, we may, if we choose, adopt the theory as a provisional hypothesis, and such a theory will be significant because capable of being further tested. Mach's arguments show exactly the reverse of what he thinks they show,[2] viz. they show that if Absolute Space did exist, one could demonstrate the existence of rotation relative to it.

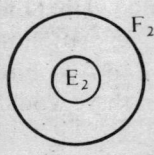

Reichenbach [9] discusses these issues with an artificial example. He imagines two world systems far apart from each other, each of an Earth (E) surrounded by a shell of stars (F) — see Figure 1. E_1 and F_2 are stationary relative to each other, and so are E_2 and F_1; but E_1 and F_1 are in relative rotation, and so are E_2 and F_2. Now, he argues, if centrifugal effects,

Figure 1

---

[1] See, e.g. K. R. Popper, *The Logic of Scientific Discovery* (London, 1959), *passim*.
[2] One could just interpret Mach in the cited passage as claiming that it was rash

such as water stationary relative to the Earth having a deformed surface, are found in both E_1 and E_2, that shows that they are produced by mere relative rotation of earth and shell, and hence we do not have to, indeed cannot, attribute centrifugal effects to absolute rotation; in our terms, E_1 and E_2 will be frames of reference relative to which the laws of nature are equally simple. But if centrifugal effects are found in only one Earth, say E_1, clearly no mere relative rotation of earth and shell can be responsible for them. Hence he concludes, as we have done, that Mach's principle that centrifugal forces are connected only with relative rotation of masses 'is undoubtedly empirical' ([9], p. 217). For observation can settle this issue.

However, Reichenbach claims that, although we must postulate absolute rotation if centrifugal effects are found only in E_1, we still cannot tell whether it is E_1 or E_2 that is rotating absolutely. For, he claims, we do not know whether centrifugal forces are produced by the absolute rotation of a body or the absolute rotation round it of a shell. If the former, then E_1 is rotating absolutely; if the latter, E_2 is rotating absolutely, but we do not know which.

But this is not so. We can perfectly well determine by other experiments whether centrifugal forces are produced by the absolute rotation of a body or by the absolute rotation of a surrounding shell. We have only to rotate a body on E_1 in the opposite direction to that in which F_1 is rotating relative to E_1 and see if greater centrifugal effects appear in it. If they do, it is the absolute rotation of a body which produces in it centrifugal effects. If they do not, then it is the rotation of the surrounding shell which produces centrifugal effects. Hence we can determine whether E_1 or E_2 is rotating absolutely. Reichenbach's example shows that if Absolute Space did exist, one could detect by the issue of experiments which bodies were rotating relative to it, and not merely that some bodies were rotating relative to it but not which bodies those were.

However, we have urged, the claim that there is Absolute Space is the claim that there is a single most basic frame and not a class of equally basic frames. And for scientific, not philosophical, reasons, such a claim is not justified. Still, if the empirical data were such that a justifiable

---

to adopt a theory that the bucket was rotating relative to Absolute Space when it rotated relative to the fixed stars and showed deformation, until we knew the result of experiments with a thickened bucket. But Mach seems to be claiming not this, but that until such experiments had been performed, the theory was not significant.

scientific distinction (in the sense earlier described) could be made between absolute velocity and mere relative velocity (and not, as discussed above, merely between absolute rotation and mere relative rotation) a Machian could always raise a similar objection to that cited earlier against the interpretation of such data. Certainly, he might admit, it is simpler today to refer the laws of nature to a frame K than to a frame K' moving in uniform motion relative to it, and to attribute certain effects found on K' but not on K to the motion of K' relative to K. But if you say on the basis of this that K is absolutely at rest, you commit yourself to the doctrine that the following experiment could be performed: stop K' so that the simple laws which held on K relative to K now also hold on K' relative to K', and then give K a uniform motion relative to K' in such a way that the same laws continue to hold on K' relative to K', and the effects will appear on K. 'But no one is competent to say how the experiment would turn out.' But in those circumstances the supporters of the doctrine of Absolute Space are predicting that the experiment could be performed and have the result described, and claim that they have some grounds for making this prediction. The opponents of the doctrine of Absolute Space must claim that at some stage the experiment will break down: either K and K' cannot be held stationary in such a way that the laws originally holding on K hold on both, or K cannot be given a uniform motion relative to K' so that the simple laws continue to hold on K', or, if such motion can be given, the effects in question will not appear on K. They must claim this for otherwise the effects in question cannot be produced by the mere relative motion of K and K'. The fact that a doctrine of Absolute Space does have such predictive consequences and so can be confirmed or disconfirmed by observation shows, even for the verificationist, that talk about Absolute Space is significant, and, on present scientific data, that the assertion of its existence is unjustified.

There is another thesis which is sometimes confused with the thesis of the existence of Absolute Space, as I have presented that thesis and as — historically — I claim that it has on the whole been discussed. On this second thesis to say that Space is absolute is to say that its properties are independent of the physical objects which it contains, whereas to say that it is relative is to say that those properties are so dependent. The properties of Space are its geometrical properties, being Euclidean or hyperbolic and such-like. I will call these theses the thesis that Space is M-absolute and the thesis that it is M-relative. Einstein certainly held the latter thesis: 'According to the general theory of relativity, the geometrical properties of space are not independent, but they are determined by matter' ([7], p. 113). And Newton in claiming that Absolute Space 'remains always similar' (words which I omitted from the definition of Absolute Space cited on p. 50 while considering the main issue) was committed to the view that Space was M-absolute, for since the density of matter and so physical objects in general varies from place to place (however they be identified) his claim implies that the geometry of Space does not depend on the density of physical objects. Which of these two theses is correct and how one could prove which was correct are questions which we must postpone until Chapter 6 where we will consider how it can be shown what is the geometry of Space. What is important here is that we should realise that the two theses are different theses from the ones with which I have been concerned for most of this chapter. Any confusion between them arises from the fact that Newton was both an absolutist and an M-absolutist and Einstein was both a relativist and an M-relativist.[1]

Clearly a man could consistently be an absolutist and an M-relativist. Aristotle could perfectly well have held that the geometry of Space varied from place to place with the density of the matter which it contained, while holding to the view that the laws of nature were simpler if referred to the Earth than if referred to any other frame. And, even more obviously, a man could be, as Leibniz clearly was, a relativist and an M-absolutist. A man might hold that there were many equibasic frames of reference while asserting that geometry remained Euclidean whatever objects occupied Space. Matter might deform measuring rods, he could argue, so that they were inaccurate measures of Space — but Space itself remained Euclidean. The two sets of theses are completely independent of each other.

A question closely connected historically with the questions whether Space is absolute or whether it is M-absolute is the question whether Space would exist if there were no physical objects occupying Space. This question is sometimes raised by asking whether, if the Universe came into existence at a certain temporal instant, there was empty Space before that instant existing from all eternity; or whether, if the Universe were destroyed, Space would continue to exist.

The connection of this question with the preceding ones is this. The absolutist and the M-absolutist tend to think of Space as an entity, not very dissimilar from the physical objects which occupy it. For the absolutist it is something relative to which physical objects may move,

---

[1] For examples of the interweaving of the different doctrines see [11] and the historical survey in [10].

and for the *M*-absolutist it has properties. Hence they tend to think of it as something independent of physical objects, and so find no reason for supposing that it necessarily came into existence with the first physical object. But if it did not come into existence then, what would it mean to say that it came into existence at some prior temporal instant? What would it mean to say that Space was created twenty thousand million years ago, but the Universe only fourteen thousand million years ago? Hence, the argument goes, it must have existed from all eternity. The relativist and the M-relativist as such have on the other hand no grounds for thinking of Space as an entity like the things which occupy it. For them 'Space' is just a useful term for talking about the relations between spatial things, so why say that there was any space before there were spatial things? Hence we find Clarke, the disciple of Newton, absolutist and *M*-absolutist, claiming that, although the Universe was created, 'Space is eternal' ([4], p. 47), and Leibniz, relativist, claiming that Space is only 'an ideal thing' ([4], p. 70), 'an order of things which exists together' ([4], p. 26).

However, the reasons which Leibniz and Clarke and others in their tradition give do not seem to be to the point, for the question of whether Space is absolute is, I have argued, an empirical question, and I shall argue that the same is true to some extent about whether Space is *M*-absolute. But there seems to be a conclusive non-empirical reason for claiming that Space would not exist if there were no physical objects occupying Space, reasons not directly connected with the issues of there being Absolute Space or Space being *M*-absolute, and so not brought out by Leibniz or others in his tradition.

The reason is as follows. As we saw in Chapter 1, talk of a place being the same place as an earlier place, is talk of it having the same spatial relations to physical objects forming a frame of reference (that is to material objects or to mere physical objects if the criteria of identity thereof could be made sufficiently precise for them to form a frame of reference). Hence if there are no physical objects by reference to which places can be picked out, (and so it makes no sense to talk of places as the same or different), there can be no continuing places and so no continuing space. Suppose the Universe to be annihilated and then after a time interval a new Universe to come into being. If there could be space without matter, then the new Universe could have come into existence either in the same space as the old Universe, or alternatively, in a different space. In the former case the new physical objects would have occupied the same places as places of the old space. But any given new place could only be the same place as some place of the old space if it preserved spatial relations to physical objects of the space; and it could not do that because there would not be any continuing physical objects. There can therefore be no space without physical objects occupying it.

This conclusion only holds if 'space' is understood in the natural and ordinary-language way in which we have analysed it so far, as a collection of places which could be occupied by material objects. In this sense, Space is something immaterial and not endowed with energy, which is a container for material objects. However there are physical theories such as geometrodynamics [15] which give a rather different meaning to 'Space'. For these theories the fundamental physical entity is 'Space–Time'. In them, Space–Time and so the spatial aspect, 'Space' is characterised at each point, not merely by geometrical properties, but by properties of energy and density; so that 'Space' has a role in physics similar to that of 'ether' in the nineteenth century. Kinks in Space correspond to ordinary material and physical objects; but Space itself is really regarded as a continuous physical object. In such a theory there could indeed be a flattened Space containing no material objects or currently recognised physical objects. But that is only because in such a theory 'Space' denotes an extended continuous physical object.

BIBLIOGRAPHY

[1] Aristotle, *On the Heavens*, Book ii, chs 13–14
[2] Galileo Galilei, *Dialogue Concerning the Two Chief World Systems* (trans. Stillman Drake), Berkeley and Los Angeles, 1962.
[3] I. Newton, *Principia*, Scholium to Definition viii.
[4] *Leibniz-Clarke Correspondence* (originally published 1717) (ed. H. G. Alexander), Manchester, 1956.
[5] G. Berkeley, *De Motu* (originally published 1721), §§ 52–66.
[6] E. Mach, *The Science of Mechanics* (originally published 1883) (trans. T. J. McCormack), La Salle, USA, 1960, ch. 2, section 6.
[7] A. Einstein, *The Theory of Relativity*. A popular exposition (trans. R. W. Lawson), London, 1920.
[8] A. Einstein *et al.*, *The Principle of Relativity*. A collection of original memoirs on the Special and General Theory of Relativity (trans. W. Perrett and G. B. Jeffery), London, 1923.
[9] H. Reichenbach, *The Philosophy of Space and Time* (originally published 1928) (trans. M. Reichenbach and J. Freund), New York, 1957, § 34.
[10] J. D. North, *The Measure of the Universe*, Oxford, 1965, ch. 16.

For modern philosophical discussion see:

[11] J. Earman, 'Who's Afraid of Absolute Space?', *Australasian Journal of Philosophy*, 1970, 48, 287–319.

[12] Clifford A. Hooker, 'The Relational Doctrines of Space and Time', *British Journal for the Philosophy of Science*, 1971, 22, 97–130.

[13] Sheldon Krimsky, 'The Multiple-World Thought Experiment and Absolute Space', *Nous*, 1972, 6, 266–73.

And for an especially clear and illuminating discussion:

[14] L. Sklar, *Space, Time, and Spacetime*, Berkeley and London, 1974, ch. 3.

On Geometrodynamics see:

[15] J. C. Graves, *The Conceptiual Foundations of Contemporary Relativity Theory*, Cambridge, Mass., 1971.

4 Distance and Direction — (i) Primary Tests

What are the criteria for saying that some object or place is at a certain distance in a certain direction from another object or place?

There are many and various ways of finding out what is the distance and direction of *A* from *B*, but the primary tests which give the meaning of distance and direction would seem to be the rigid rod tests which I shall shortly describe.

ϕ is a primary test for some property, e.g. the distance between two points, if it is a member of a set of tests ϕ, χ, ψ, etc., such that if the use of all those tests gives one definite value for the property, then the value given by those tests is the value of the property. No different tests can upset the common result of all the primary tests. α is a secondary test for a property, if we only use it because we believe that it will give the same result as one or more primary tests. If a secondary test gives a result different from that given by the primary tests, the former is considered wrong and the latter right. If different primary tests give in some particular situations different results from each other, then the concept of the property in question cannot be unambiguously applied in the situation. There is no clear answer to the question, what is the value of the property in that situation. If in general different primary tests give different results then the concept of the property in question is no longer of use; we need a new concept. If in general different primary tests for distance, weight, or temperature gave different results, talk about the distance between two objects, the weight or temperature of an object would be useless. The primary tests for distance and direction which I shall describe are highly cumbersome and often practically impossible to apply. We do in practice normally use other tests, and I shall later substantiate my claim that these other tests are secondary tests.

My aim in describing the primary tests for distance and direction is to formalise, make precise, our ordinary concepts of distance and direction. I claim that there are perfectly clear primary tests for distance and direction, which give definite results to a high degree of accuracy in ordinary situations on Earth. However, it is not always clear what are the tests to be used to give results to an even higher degree of accuracy, nor how the tests are to be applied when we are concerned with regions of Space very different in character from the region near the surface of the Earth. I shall show that sometimes there is a clear way to use the tests in extraordinary situations, but that sometimes it is a matter for arbitrary decision how we are to use them there.

The primary tests for distance from one material object A to another one B are as follows. We define a certain specified rigid rod (subject to certain conditions to be described later) to be one unit of distance. The most common unit of small-scale distance is the metre, and the standard metre is to be found in Paris. A rigid rod is a rigid body whose height and breadth are small in comparison with its length (tests for a body being a rigid body will be discussed on pp. 69ff). Any other rigid rod which coincides in length with the standard rod is also said to be one unit long — both when it is where the standard rod is and anywhere else. The distance from A to B is then given by the smallest number of times we need to place such a unit rod along a line from A and B in order to reach B. Distance along a line is measured as follows. Two points are marked on the rod. We place one against A and mark the point where the other touches the line. Then we place the first against the mark made by the second and continue this process until we reach B. The line (which will normally be marked on a surface) must be of constant length, viz. neither expanding nor contracting and stationary relative to the starting point A. The test that it is of constant length is that the distance between any two points on it remains constant over time. If we have placed the rod along the line n times in order to reach B, then the distance from A to B along the line is n units. Measurements of distance made with a rod, the length of which is tailored to the length of different standard rods (e.g. the standard metre and the standard yard), may be said to be made in different units but on the same distance scale, since measurements made in one kind of unit are a constant multiple of measurements made in the other kind of unit. This latter condition, we will see later, must be satisfied if both standard rods are rigid bodies.

Distances along lines on to which a rigid measuring rod of standard length cannot be fitted an integral number of times are measured as follows. If the distance to be measured D is longer than that measured by laying down the rod n times but shorter than that measured by laying down the rod $n + 1$ times, then D is between n and $n + 1$ units of length long. The fraction F by which D exceeds n units is then measured by the following procedure. A rigid rod of length F is constructed and integers k

and *m* found such that the rod can be fitted exactly *k* times against *m* rigid rods of unit length along a straight line. *F* will then have been shown to be $\frac{m}{k}$ units of length long. If *m* and *k* can be taken as large or as small as we please we can measure to any degree of accuracy we wish. However there may be physical limitations on the length of rods preventing *m* and *k* being taken as large or as small as we please. The physical limitations, to which we called attention on p. 27, are that it may not be physically possible to measure differences in length of less than a certain minimum amount.

There seem to be at least two different primary tests for a line being a straight line,[1] (and possibly other tests of a similar mechanical character, which I have not elucidated). The first test is the shortest-distance test. By it a line is a straight line if the distance between any two points on the line is shorter than the distance between those points measured along any other line close to that line which could be constructed joining them. (If two neighbouring lines joining two points were of equal length but shorter than other lines close to them joining the same two points, it seems unclear whether we ought to say that both are straight lines or that neither are. Usage here is vague and needs tightening if we require clarity.) We cannot of course measure the distance along all lines which could be constructed joining all points, but we can have good inductive evidence that a certain line is a straight line. This would be provided by finding for various pairs of points on it that lines joining them on either side of it were longer than it and that the lines beyond them longer than them and so on. The lines referred to are lines of constant length. We can use this test to provide a primary test for a plane. The test that a surface of constant size is a plane surface is that all pairs of points on it are joined by straight lines lying along it. An inductive test similar to that needed to show a line to be a straight line can be performed to show that the primary test is satisfied. The test that a surface is of constant size, that is, neither expanding nor contracting, is that the distance on straight lines between any two points on it remains constant over time.

---

[1] These tests for a line being a straight line do not presuppose the truth of any one type of geometry. They are compatible with Euclidean or hyperbolic geometry or geometry of any other type. Sometimes geometers only use the term 'straight line' where the geometry is assumed to be Euclidean. Ordinarily however, even if we knew nothing about geometry, we would seem to be able to ascertain that a line is a straight line. Hence I make no assumption about geometry in using the term 'straight line'.

The other primary test for a line being a straight line we may term the three-surface test, since it begins with a three-surface test for a plane. If a surface *A* fits snugly on another surface *B*, and *B* fits snugly on a third surface *C*, and *C* fits snugly on *A* (such that for all points α on *A* touching a point β on *B*, the point β touching a point γ on *C*, then when *A* fits snugly on *C* the points α and γ fit on each other) then *A*, *B* and *C* are all planes. Then if two planes have a common edge, that edge is a straight line. For practical reasons the second test can only be used over very short distances, but where both tests can be used, we expect the tests to give the same results as each other. If they do, then the line that by both is a straight line is really a straight line and the surface that by both is a plane is really a plane. The description of the tests elucidates our ordinary concepts of 'straight line' and 'plane'. If in one instance the two tests gave different results, we whould not know in that instance how to apply our concepts. If frequently whatever satisfied one test did not satisfy the other and conversely, then our ordinary concepts would have proved useless. We would have to modify them, so that perhaps satisfaction of one test alone sufficed for a line to be termed a straight line. There is no logical necessity that the tests should give the same results, but empirically we have always found, and all modern theories of mechanics predict that we will always find, that when the one test is satisfied, the other is too. Similar considerations would apply if there are further primary tests for a line being a straight line.

To measure the direction of *B* from *A* we have to measure the angles which a straight line from *A* to *B* makes with other straight lines which pass through *A*. Normally measuring the angles made by the straight line from *A* to *B* with any two straight lines will suffice, with measurement of the distance along the former line, uniquely to identify *B* (with peculiar geometries or in spaces of dimensions greater than three, if such there could be, more angular measurements would be needed — for this see Chapter 7). Measurement of the necessary number of angles gives the direction of *B*. To measure the angles, we construct a protractor. We mark a straight line on the plane surface of a rigid body. Along this line we lay a thin rigid rod rigidly connected at one end to another thin rigid straight rod. The test that the rods are straight is that they fit along straight lines. The test of rigid connection is that the distance between any two points on the two rods remains constant as the rod is moved. The rods are then rotated in the plane of the surface about the point where they are joined. The angle between the two rods in their original position is given some arbitrary value. The angle between the rods in each subsequent position which they take up is then said to have the same value. There are 360 degrees in a complete circle. So when the rigid rods are so joined that 360 of the angles between them make the whole circle, then the angle between them will be one degree. We mark off the divisions by straight lines meeting at a centre point on the plane surface which we now call a protractor. We can then measure the angle between any two intersecting straight lines by fitting the protractor on to them so that one lies along its base line and the other coincides with another line of it. The number of unit angles marked on the protractor which fit between the two straight lines gives the angle between the two lines. It is conceivable that there may be more than one direction of B from A. This will arise if more than one straight line goes from A to B — which will not however occur if the geometry of Space is Euclidean or of many other types. The applicability of the concept of direction pressupposes the applicability of the concept of a straight line. Given the latter, then any two points which can be connected by a line (and so are at some distance from each other) can be connected by a straight line, for the shortest line will be the straight line. Hence any two spatially related points will lie in some direction from each other.

The distance and direction established above is the distance and direction from A to B at the temporal instant at which the rod placed on the straight line stationary relative to the starting point touches B. (How to ascertain which instant at A is simultaneous with this instant is a question which we shall consider in Chapter 11.)

To apply the criteria described we need to be able to recognise coincidences (of rod and mark) which we can do either visually or tactually. It should be noted that even if we rely on visual means to detect coincidences, we do not thereby pre-suppose the rectilinear propagation of light. Even if light travelled in circles, a coincidence would remain a coincidence.

The methods for measuring distance and direction which I have described are methods for measuring the distance between material objects (e.g. the distance from this wall to that wall). They may be used for ascertaining the distance between other spatial things, in so far as the boundaries of those things are distinct enough for distance to be measured. But places as well as spatial things have from each other distance and direction. To measure the distance between where A is and where B was is to measure the distance between two places. In order to identify the present place which is the same as the place previously occupied by B, we have to specify or take for granted a frame of reference.[1] The distance between two places P and Q is naturally supposed to be the distance between two objects occupying those places.

At this point difficulties arise in applying the ordinary concept of distance. It is a consequence of the Special Theory of Relativity, and of all more general theories of mechanics which include it, that the distance so measured between two places depends on the velocity of the objects occupying those places — and this, whatever standard of simultaneity is adopted. Thus suppose that two places on an inertial frame P and Q are occupied at the same instant, as judged by some standard of simultaneity, respectively by two objects a and b stationary relative to each other and that the distance from P to Q as measured by the distance from a to b is x. Suppose another object a' to be at virtually the same place as a at that instant and to be passing a with a velocity v away from b. The distance from P to Q as measured by the distance from a' to b will then be different from x — whatever our standard of simultaneity. Suppose that the standard is the light signal method (to be described in Chapter 11) so that clocks on a and b have been synchronised by light signals sent between them. Then the distance from a' to b will be — according to special theory — a longer distance βx where $\beta = \dfrac{1}{\sqrt{1 - \dfrac{v^2}{c^2}}}$ (c being the velocity of light in any inertial frame). Hence the distance between two places will depend on the velocity of the objects occupying those places, between which distance is measured. This difficulty about the distance of one place from another makes awkward the previous analysis of the distance of one material object from another. We measured the distance from one

---

[1] In the article referred to on p. 25 D. M. Armstrong argues that it is implausible to suppose that we need to presuppose a frame of reference in order to give a value to the distance between where A is and where B was or will be; and that this implausibility lends support to the view that there is Absolute Space. For, he argues, if there is Absolute Space, there will be a unique way of reidentifying places (viz., by a most basic frame) and hence of measuring distances between objects at different temporal instants. Hence, he claims, the expressions 'distance between where A is and B was' or 'distance between where A is and B will be' will not be elliptical. But even if one believes that there is an Absolute Space, one still has to specify the frame of reference by which one is reidentifying places; for, unless the Earth is still in Absolute Space one does not normally use the most basic frame for reidentifying places. The allegedly implausible doctrine of measurement of distance is evidently true. The distance between where the driver of the car now is and where the picnic basket was depends on the frame of reference which you choose for reidentifying places.

material object *a* to another one *b* in the frame in which *a* was at rest because the distance shows how far it is for an observer situated at *a* to get to *b*. But the distance from *a* to *b* is the distance from the place where *a* is to the place where *b* is, and what this distance is depends on the frame in which we measure it.

For these reasons expositions of the Special Theory of Relativity and more general theories of mechanics and cosmology normally substitute for the concept of distance the concept of distance in a frame. Thus we can distinguish the distance from *P* to *Q* in the frame *F* in which *a* is at rest, the distance in the frame *F'* in which *a'* is at rest, and so on. The distance from one object *a* to another one *b* in the frame in which *a* is at rest is the distance measured as previously described. The distance at a certain instant from *a* to *b* in the frame *F'* in which *a'* is at rest, is the distance in *F'* from *P*, the place on *F'* occupied by *a* at that instant, to *Q*, the place on *F'* occupied by *b* at the same instant. The distance from *P* to *Q* in *F'* is the distance from an object at *P* stationary in *F'* to an object at *Q*.

Now certainly the frame often need not be explicitly stated, especially in ordinary discourse about mundane matters. It may be obvious in which frame distance is expected to be measured. By the distance from one material object to another is normally meant distance in the frame in which the first is stationary. Thus when the astronomer talks of the distance from the Earth to a distant galaxy he means the distance as measured in the frame in which the Earth is stationary. The distance measured in the frame in which the first object is stationary is termed in relativity theory the 'proper distance' from *a* to *b*. Likewise one frame of reference is normally the obvious one in which to measure the distance from one place to another. This is because, as we saw in Chapter 1, any place must be identified by reference to a frame. Hence we normally understand by 'the distance from *P* to *Q*' 'the distance from *P* to *Q* in the frame by which *P* is identified'. For only in this frame will *P* be stationary, and hence the process of measurement be able to be carried out by measuring the distance at all instants of other places from *P* and then noting that at a certain instant *Q* was at one of them. If we wished to measure the distance in another frame, we would have to ensure that the instant at which we marked a point on the frame at *Q* was the same as that at which we marked our point at *P*. Unless the distance from *P* to *Q* is measured in the frame at which *P* is at rest, we thus need to have and apply a standard of simultaneity before we can ascertain what the distance is at any instant at Q. For these reasons there is normally a natural interpretation of 'the distance from *P* to *Q*'.

Difficulties arise when a place is specied, as we noted (p. 12) that it may be at some temporal instant, by more than one frame of reference. Suppose two sticks *a* and *b* to be erected on Earth ten feet apart and a knob *a'* to portrude from a space rocket passing over the Earth. Suppose *a* and *a'* to coincide at some instant. Then what is the distance between the place occupied at that instant by *a* and *a'* and that occupied by *b*? Here indeed 'distance' is elliptical and we need explicit specification of the frame in which it is to be measured before we can say how far apart are the places.

Nevertheless although different measurements of distance between places result from different choices of frame in which to measure, as we saw in the last chapter, for accurate formulation of the laws of science, places ought to be reidentified by the most basic frames of reference that there are. It would follow that distances from one place to another ought to be measured relative to such frames. If there is only one most basic frame of reference, then there will be a unique way of measuring distance between places which science ought to adopt. Yet, as we saw in the last chapter, given the truth and universal applicability of the Special Theory of Relativity, each of infinitely many inertial frames are equibasic frames of reference. However, once we pass beyond Special Relativity to cosmological theories of the Universe occupied by homogeneously distributed matter (the meaning and justification of which we shall consider in Chapter 11) we find that in the region of each galactic cluster, there is (see p. 51) one most basic frame and hence distance ought to be measured relative to it. On the cosmic scale, however, each of very many fundamental particles in relative motion is equibasic, and hence science can choose between many frames for its measurements of distance. If the Universe proved not to be homogeneous, whether or not there is a unique frame in which distance ought to be measured for the purpose of formulating scientific laws depends on whether the scientific evidence is that there is a most basic frame of reference. Similar difficulties and solutions arise with direction.

There is no logical necessity that at some instant the distance from *A* to *B*, these being material objects or places, be the same as the distance from *B* to *A*. The distance from *A* to *B* is measured at *B* along a plane on which *A* is stationary, and the distance from *B* to *A* is measured at *A* along a plane on which *B* is stationary. If *A* and *B* are stationary relative to each other, the two distances will be the same since the line along which the distances are measured will be the same. Otherwise, whether or not at any instant these are the same distances, will depend on our criteria for simultaneity of instants at *A* and at *B*. Clearly the difference between the two distances will only be large if the velocity of mutual recession or approach of *A* and *B* is large. I shall argue however in Chapter 11 that the criteria of simultaneity on the cosmic scale are such that where *A* and *B* are clusters (which, if clusters far distant from each other, will have enormous velocities of mutual recession) that the distance from *A* to *B* will be approximately the same as the distance from *B* to *A*. In so far as the distance from *A* to *B* is virtually the same as the distance from *B* to *A*, we can refer to either as 'the distance between *A* and *B*'. But when the two former distances are not virtually the same, we need to make clear which we mean by 'the distance between *A* and *B*' before its value can be established.

So in this respect the ordinary concept of distance has in the light of scientific discoveries to be replaced by the concept of distance in a frame. Yet as it is normally obvious in what frame the distance is to be measured, it is not normally necessary explicitly to state it, and in subsequent discussion of distance it will not normally be necessary for me to do so.

The tests described involve using a rigid rod. When we say that a rod is rigid, we mean that it does not alter its length, viz. remains congruent with itself, under transport or when environmental conditions change in various ways. But how can we show that the rod does not alter its length in these ways? We cannot measure it to show that it has not changed its length, for it is itself our standard of measurement. So what criteria have we for rigidity?

There are two criteria which we ordinarily use for the rigidity of a material object. First, given that — as is in fact the case — there is in the Universe one and only one family of material objects preserving approximate coincidences among themselves, then these are approximately rigid bodies. The family must include many bodies constructed out of many different materials. The bodies must be such that if they coincide in length with each other at one place in one set of environmental circumstances, then they will coincide approximately with each other at very many other different places under very many other different environmental conditions (and so if they do not approximately coincide in length at some one place, they will hardly ever approximately coincide). Bodies of jelly, plasticine or rubber, if they coincide at *P*, will not in general coincide at *Q*. Bodies of wood, iron, stone, etc. if they coincide at *P* will in general coincide at *Q*. Consequently by this criterion we may say that — approximately — they are rigid. True, there are some conditions, such as conditions of very high temperature, under which almost all coincidences are destroyed.

But all that is necessary for the satisfaction of this criterion is that under many diverse conditions coincidences under transport should be preserved approximately.

This first criterion is a condition for approximate rigidity. If there are material objects which satisfy it, they and no other objects are the candidates for rigid bodies to be assessed by the second criterion. It was the fact that there were bodies satisfying the first criterion, bodies such as rods of wood or bronze or iron, which no doubt led to the emergence of the concept of rigidity and so of the concept of length. However, though various bodies satisfy the first criterion, in the sense that they coincide approximately in different situations, there is no family of bodies coinciding exactly in many different situations. This suggests that most of these bodies are subject to slight changes of length with changes of temperature, pressure, etc. The problem then is to find the extent of these changes. Scientists then adopt that hypothesis about slight changes of length caused by deforming forces which leads to the most accurate satisfaction of the simplest scientific laws. This is the second criterion for rigidity. Those bodies which satisfy the first criterion are to be assumed rigid in so far as and to the extent to which this assumption leads to the simplest scientific theory giving accurate predictions governing changes of length with temperature, pressure, etc.

Now the second criterion cannot be applied until the first has been satisfied. That celebrated theory that 'all bodies expand uniformly with time' might turn out to be a very simple theory explaining a vast number of phenomena better than existing theories. By this standard the only body which remained 'rigid' for even a small temporal period would be something like a tube of sand in a sand clock. But any theory which showed that the sand in a sand clock was really a 'rigid' body, whereas rods of iron and steel were not in the least rigid, would be telling us something about 'rigid' bodies but not about rigid bodies. The result is so contrary to what can be observed straight off that it is clear that 'rigid' is being used in an unusual sense. Iron bodies and stone bodies are paradigm cases of approximately rigid bodies, which supports the view that the first criterion must be satisfied by any candidate for being a rigid body.

Let us now consider how the second criterion works. Deformations are normally detected by rods which previously coincided exactly no longer coinciding exactly. Suppose two thin rods, I made of invar and L made of lead, coincide exactly at P. When taken to Q, they only approximately coincide, L being slightly longer. We seek the reason for the difference and note the higher temperature at Q. We also notice that I

and *L* coincide again exactly at *R* where the temperature is the same as at *P*. The simplest hypothesis about the difference at *Q* appears to be that it is due to the higher temperature. But higher temperature may mean that *I* and *L* have both expanded, but *L* more than *I*; or that *I* and *L* have both contracted, but *I* more than *L*; or that *I* has contracted while *L* has expanded. How can we test between these alternative forms of the hypothesis that it is the higher temperature at *P* which was responsible for the difference in length of *I* and *L*? We must heat *L*, and then bring it, at *P*, into momentary contact with *I* and see which is the longer. The contact must only be momentary because otherwise the temperature of *I* will also be affected. We note that *L* is longer. If we heat *I* and bring it into momentary contact with *L*, we find that *I* is longer. Hence, assuming the same forces to operate at *P* as at *P'*, viz. making the simplest assumption about the forces operative, we can see that the difference in length is due to the expansion of both *I* and *L*, and the extent of the expansion can be measured by measuring at *P* the effect of an increase of temperature on each body separately, the other being kept at the original temperature.

Other deforming influences (e.g. pressure and magnetic field) can also be detected by noting slight differences of length in different environments. Once we have ascertained how bodies are affected by temperature and other environmental factors, we find that no rod is perfectly rigid under many circumstances. Any rod will change its length slightly when moved around the surface of the Earth. However, once we have ascertained what are the deforming influences and their exact effect we can infer what distance between two points would be measured by an ideally rigid rod. We can find this by correcting actual rods which are approximately rigid for deforming influences. Statements about distance are statements about the behaviour of ideally rigid rods, but their truth can be ascertained by observing the behaviour of approximately rigid rods and correcting the results by the laws of deformation.

Another way of finding deforming influences is to use some physical process which can be repeated in each different environment and which yields correlations with length measurement. The normal one taken is the wave-length of light. We find that the wave-length of a given type of light calculated in many environments by interference fringes is uniform to a very high degree of approximation, if measurements of the distances from which the wave-length is calculated (distances from sources to fringe, and between fringes in an interference experiment for determining wave-length) are made by rods corrected by the methods previously described for known deforming processes. The simplest hypothesis to make is that the wave-length of any given type of light is uniform in all environments and that any apparent differences are due (as well as to errors of calculation) to hitherto unknown influences deforming the rods measuring the distances from which the wave-length is calculated.

So in these ways we account for slight differences in length between approximately rigid rods by postulating the simplest laws of deformation which will account for them. Now at this point a circularity seems to arise. Clearly the laws of deformation have to be formulated on the assumption of some geometry, for they make statements about inferred distances (viz. distances which cannot in practice be measured directly — e.g. wave-length of light), areas and volumes, which can only be calculated from measured distances when we know the geometry of space. Thus if we are to apply a correction to our rods for temperature change, we must be able to measure temperature. We can do this because the volume of substances (e.g. mercury) varies in a regular way with their temperature. So to measure temperature we must be able to measure volume. We can calculate volume from measurements of length (e.g. the volume of mercury in a test tube by measuring the length of the column and the diameter of the tube) if we have a formula for calculating the volume of things from distance measurements. But such a formula will vary with the geometry — Euclidean geometry yields a different formula for this from hyperbolic geometry. Geometry states what propositions about area and volume, and further propositions about distance and direction, are entailed by given propositions about distance and direction.[1] We cannot therefore formulate the laws of deformation until we have ascertained the geometry. But we cannot ascertain the geometry until we have made some measurements of distance and direction — given what we shall demonstrate in Chapter 6, that the geometry of space is a logically contingent matter. Only then can we see what propositions about distance and direction entail other such propositions and hence infer propositions about area and volume. But we cannot make measurements of distance and direction until we know how to correct the results of measurements made by approximately rigid bodies. And we cannot make these corrections until we have formulated the laws of deformation — so where do we start?

What we must look for is a value of G_r, the geometry resulting from measurements, for which $G_r = G_a$, G_a being the geometry used in formulating the correcting laws.[2] The trouble is that there might be no

---

[1] For detailed exposition of these points see Chapter 6.
[2] This was argued by Grunbaum [2] pp. 144–7.

value of G_r for which $G_r = G_a$, or, alternatively, there might be more than one value.[1] To claim the former is to claim that any formulation of laws of deformation consistent with phenomena on the assumption of some geometry G_1 leads us when measuring space, to find that the geometry was G_2 where $G_1 \neq G_2$. Thus the process might be cyclical. If we assumed Euclidean geometry in formulating our correcting laws, and corrected our measurements in the light of them, we might discover that Space had a certain non-Euclidean geometry. We might incorporate this result into our laws and then find by measuring with these laws that Space is Euclidean and so *ad infinitum*.

But then while we assume any one geometry, there may be one simplest set of laws of deformation to which we can extrapolate from recorded phenomena of change of relative length with temperature, etc., there are an infinite number of less simple sets of laws. We should have to resort to less simple laws to cope with this difficulty. The simplest set of laws for a finite collection of observed phenomena is, we have seen (p. 43), the one which on balance uses the simpler mathematics or other symbolism and forms the more coherent system; and, for any body of phenomena, as we have seen, we ought to choose the simplest set of laws. If our phenomena are not merely the phenomena of deformation but also the other phenomena for which our theory of physical geometry is constructed, we are required to choose the simplest set of laws compatible with both collections of phenomena. Such a set may mean that the set of laws applicable to either collection of phenomena taken in isolation will not be the simplest set that could be provided for it. Nevertheless it is a paramount requirement that scientific theory be self-consistent. If the choice of less simple laws of deformation (e.g. about the way in which length changes with temperature), enables us to choose a geometry consistent with physical laws, then this requirement of making science as a whole (physical geometry and physical laws) self-consistent demands our adopting the less simple hypothesis. The supposition that, with an infinite number of possible sets of laws of deformation formulated subject to the assumption of any geometry G_a, no value of $G_a = G_r$ could be found seems unreasonable. But if, as is unlikely, this situation did arise, we would have to say that the ordinary concept of distance proved incoherent and could not be applied rigorously in describing the physical world.

A much more likely situation to arise is that more than one solution be

[1] See Fine [4].

found to the equation $G_a = G_r$. That is, if the simplest possible laws of deformation are formulated on the assumption that geometry is G_1, measurements made with the aid of rods corrected by these laws show that the geometry of Space is G_1. But also if the simplest possible laws of deformation are formulated on the assumption that G_2 is the true geometry, measurements show that it is. In this situation if one of the sets of laws and its resulting geometry taken together was simpler than the other, we would adopt it. But if the two sets of laws and their geometries taken together were equally simple, we would be faced with two equally satisfactory criteria for distance. Then the ordinary concept of distance would be shown to permit of two possible rigorous interpretations, and an arbitrary choice would be needed between them.

My argument so far has been that, given that — as in fact the case — there is one and only one family of (approximately) coincidence-preserving rods, they are (approximately) rigid rods. But what would we say if there were no such family or more than one family? If there were no such family, the second criterion of simplicity would initially have to be used alone for deciding which actual bodies were most nearly rigid bodies and what corrections have to be made to the results yielded by them to give the measurements which would be made by ideally rigid bodies. It would be necessary that the resulting physical theory have the consequence that if there were any family of (approximately) coincidence-preserving objects they would be approximately rigid bodies. However, it is to be doubted whether men would have such concepts as rigidity and length in a world without a family of (approximately) coincidence-preserving objects.

What next if there were more than one family of (approximately) coincidence-preserving objects? Suppose that objects of certain materials approximately preserved coincidences under transport, and that objects of other materials also approximately preserved coincidences under transport, yet members of one family did not preserve approximate coincidences with members of the other. Here we have two potential alternative sets of approximately rigid bodies — which are we to choose? If one family contained bodies of many more different types of material than the other, or bodies which were far more abundant in the Universe than those of the other, we should be reasonably inclined on grounds of simplicity to consider bodies of the former family the true rigid bodies. Alternatively, physical theory might make possible a clear test between the rival claimants. If the physics based on the corrected measurements given by one set of bodies was far simpler than that based on the corrected measurements given by the other set, that would be good reason for taking the first set of bodies as the true rigid bodies. But in so far as there was not much to choose between the two families on these grounds, if we still decided to call members of one family rigid, the decision would be arbitrary. It is a fortunate feature of our universe that there is one and only one family of (approximately) coincidence-preserving bodies. Again, it is doubtful if we should have the concepts of rigidity or length in a universe in which there was more than one family of coincidence-preserving bodies.

One further point must be made to tidy up the argument to date. Once we have discovered the forces deforming bodies, we can specify more exactly than heretofore the conditions under which the original standard rod is to be taken as truly standard. If we find that, for example, temperature, pressure, and magnetic field are deforming forces, we must specify the temperature, pressure, and magnetic field for which the standard rod is truly standard.

The standard rod might nevertheless for unknown reasons change its length in the course of time. The evidence for this could be of two kinds. First, that all the other rigid rods which coincided previously with it when corrected for deformations, while continuing to preserve coincidences among themselves, no longer coincided with the standard rod. Then we would be forced to postulate either that all the other rods had been subject to a uniform expansion or contraction, or that the standard rod had been subject to expansion or contraction. Clearly simplicity dictates the latter choice.

Secondly, the evidence could be that we were forced to introduce considerable complexities into our physical theory for other reasons unless we supposed that the standard rod had changed its length. As we have noted, the most valuable method of checking deformations is by assuming the wave-length of any given type of light to be constant in different environments. If we take as undeformed a rod which when measured by wave-lengths of a given type of light, preserves its length, the laws of deformation adopt the simplest known form. Hence we can show that the standard rod changes its length by showing that its length does not remain constant when measured by wave-lengths of a given type of light. To use wave-lengths of light to check that the standard rod preserves its length is to adopt a wave-length standard of light. This is what physicists did in 1960 when, fearful that unknown deforming influences might alter the distance between the two scratches on the standard metre bar in Paris, they adopted the wave-length standard of light, and defined one metre as 1,650,763·73 wave-lengths *in vacuo* of radiation emitted by the transit between $2p_{10}$ and $5d_5$ of $^{86}_{36}$ Kr, in other words of an orange line of a certain isotope of krypton.

It would be grossly misleading to describe this change of standard by saying that physicists no longer understand distance in terms of rods but in terms of wave-lengths of light. For the krypton standard was only adopted because the results given by it coincided to such a high degree of approximation with those given by rigid rods corrected in the other ways described earlier for deforming influences. Further, if these methods ceased to give at all similar results, we would surely give up the krypton standard. Suppose that by the krypton standard it turned out that all iron rods moved from one side of my room to the other halved in length. We would be faced with a choice between abandoning the krypton standard, saying that the wave-length of light varied in different parts of the room, and saying that the rods contracted to half their length in transport. Since iron rods are a paradigm case of something that preserves its length approximately under transport, at any rate near the surface of the Earth, I suggest that we would abandon the krypton standard. If this is so, then 'distance' cannot really mean something determined by the krypton standard. This is not to deny that in a hundred years' time 'distance' may really mean something determined by the krypton standard. But what I am investigating is the meaning of the term as now used. The role of the krypton standard is not to replace the rod standard but to serve as the ultimate criterion of when and to what extent any rod including the standard rod is subject to *slight* distortions. It is a correcting and not a defining standard.

Now so far I have supposed that we judge that a rod has changed its length under transport if it no longer preserves exact coincidence with rods which previously coincided exactly with it, or is shown to have done so by diverging from a measurement given by some physical process (e.g. by the wave-length of light of some type assumed constant in all environments) with which it previously coincided. What of the possibility that all rods of the same length might change their length under transport in the same proportion and that such physical properties as the wave-length of krypton light be equally affected so that coincidences are preserved? If measuring instruments are all so affected that coincidences remain coincidences, then I shall say that they are subject to a universal influence. This would be so if the distance between all points on all rods was doubled on their being moved to some region. Since all rods would be equally affected, one could not show the increase of size by making measurements. If the wave-length of all kinds of light was doubled too, one could not use this means to check the increase. If on the other hand, measuring instruments of different size or construction are so affected that ones which previously coincided no longer coincide, I shall say that they are subject to a differential influence. Clearly heat is a differential influence because it expands bodies made of some materials more than others.

My definitions of universal and differential influences differ in two respects from similar definitions given by Reichenbach. Reichenbach distinguished between universal forces, these being forces which 'affect all materials in the same way', and against which 'there are no insulating walls' ([1], p. 13), and differential forces, these being forces which affect different materials in different ways and from which bodies can be protected by insulating walls.[1] There are two respects in which my distinction differs from Reichenbach's. First it is concerned with changes of length, however brought about, whether by a force or as a natural process. A force which made all bodies expand in the same proportion would be a universal influence; but all bodies might expand in the same proportion naturally, just as they may change their position, without the operation of a force.

Secondly, there might be influences which affect all materials in the same way and against which there are no insulating walls, but which affected bodies of different shape and size in such different ways that the differences would be detected by coincidences no longer being preserved. Such influences, if forces, would be universal forces on Reichenbach's definition but would not be, in my sense, universal influences.[2] Interestingly, an example of a scientific theory which postulates influences of this type is the General Theory of Relativity, if it be interpreted in its original form as claiming that near massive bodies all bodies, including measuring rods, are deformed. The second interpretation of the General Theory of Relativity and that normally adopted by physicists is that near massive bodies the geometry of Space is altered. This interpretation we shall discuss subsequently.

In the original interpretation the General Theory claims that a rod of whatever material made, which, when very very far from all other bodies has a length l_0, will, when at a distance r from a body of mass m at an

---

[1] For some of the difficulties in this definition and analysis of what it does not include as a universal force see [2], ch. 3, section A, and Brian Ellis 'Universal and Differential Forces', *British Journal for the Philosophy of Science* (1963) 14, 177–94.

[2] Reichenbach himself made the point that his concept 'of universal force is more general and contains the concept of the coincidence preserving force as a special case' — [1] p. 27.

angle ϕ to the radial direction, contract to the length $l_0\left(1 - C\dfrac{m}{r}\cos^2\phi\right)$

where C is a very small constant. Using the CGS (centimetre, grams, second) system C is 3.7×10^{-29}. The contraction is thus greatest when the rod is in a radial position (since $\cos o° = 1$ is the maximum value of $\cos\phi$), and non-existent when the rod is in a tangential position (since $\cos 90° = 0$). The rod is supposed to be small and straight and to have negligible dimensions other than length. A direct test of this claim cannot of course be made. If we measure the length of the rod in a radial and in a tangential position, the length will be the same; but this according to the theory is because the measuring rod contracts as well as the measured one.

However, the deforming force could be detected by change of relative length, for example as follows. Construct in empty Space two rods of the same material, one PQR in the form of a segment of a circle, and another $P'R'$ a straight line, such that in empty space when P and P' coincide, R and R' can be made to coincide. Suppose that the two rods preserve coincidences everywhere in empty Space and seem to be, by the criteria described earlier, rigid rods, alternative rods for measuring the same distance PR. Now hold P and P' coincident and let R' rest at R, free to

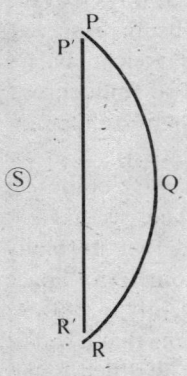

Figure 2

expand or contract. Bring the rods into the presence of a massive body S, so that S lies at the centre of the circle of which PQR is a segment (see Figure 2). Now according to General Theory PQR will be subject to no contraction since it lies everywhere perpendicular to radii stretching from S. $P'R'$ will, however, be subject to contraction and R' will retreat from R. What applies to this case applies generally. Rods of different length, thickness, and shape will change their length, thickness, and shape relative to each other in the presence of massive bodies. Coincidences in empty Space will no longer in general be preserved in the presence of massive bodies.

If coincidences between points on two bodies are not preserved, one or other body must have been deformed, so that the distance between two points on it is not the same as it was before. Differential influences, evidenced by coincidences not being preserved, cannot be 'transformed away'; a true scientific theory of deformation must recognise them. Einstein, however, suggested that we could take 'one and the same rod, independently of its place and orientation, as a realisation of the same interval' ([7] p. 161) and suppose that the presence of massive bodies did not deform bodies, but only deformed the geometry of Space. This reinterpretation of General Theory has in general been adopted by physicists following Einstein's hint. But, without further amplification, the reinterpretation is not a possible one. As we have seen, some rods must be deformed by transport. The reinterpretation will only be possible if it specifies the size and shape of rods which are not deformed by transport. We can say, if we want, that straight rods one foot long and one inch thick are not deformed, whereas all other rods are deformed. But to say this of course seems highly arbitrary. Why should rods of just this type be exempt from deforming influences? Einstein was concerned with very small straight rods of negligible thickness. The smaller and thinner the straight rod by which we measure Space, he would claim, the more accurate our measurements would be. This seems a less arbitrary claim. The grounds for adopting it would be the simplicity of the resulting physical theory.

I have earlier argued that to suppose any general and radical change under transport in the lengths of bodies which by the first criterion are approximately rigid, such as iron and stone, would violate our ordinary understanding of length. These, because they preserve approximate coincidences among themselves, give us our standard of length. It is part of the meaning of distance that distances between points and especially distances on Earth, are compared by the number of times rods which preserve approximate coincidences can be fitted between them. If we do not stick by and large by this in applications of the term 'distance' then the term has changed its meaning. The scientist is of course perfectly entitled to postulate general and radical changes of some value '*l*' of bodies of iron and stone under transport, but he is not entitled to describe '*l*' as the length of those bodies. But he might, I suggest, without violating our ordinary concept of distance, postulate, if he had reason on grounds of simplicity to do so, a small or even, in a limited region far from the surface of the Earth, significant change of length with transport of all or almost all bodies which satisfy the first criterion for rigidity. As both interpretations of General Relativity only suppose that deformations are very slight, even far from the surface of the Earth, they satisfy this demand.

The same considerations apply if a physical theory were to suppose the operation of universal influences, the lengths of all bodies being changed in such a way that coincidences were preserved. So far we have only considered corrections to estimates of distances which involved differential influences. Particular experimental results force this sort of correction on physics. But what of the possibility of universal influences producing changes of length in such a way that coincidences are always preserved? No single experimental result could force the physicist to postulate such an influence, but a theory which postulated such an influence might prove to be simpler overall than a theory which did not. Under these circumstances the physicist would surely be right to accept it, so long again as it did not violate too much our ordinary understanding of length.

The general argument of this chapter so far has been that the ancient concept of distance has to be replaced by the concept of distance in a frame, although it is normally obvious in which frame distance is to be measured and so the frame does not need specifying. With this limitation there are perfectly clear tests for distance and direction near the surface of the Earth, which give results to a reasonable degree of approximation. We have found out what these tests are by analysing the ordinary use of the terms 'distance' and 'direction', by considering what we would say about the distance and direction of various bodies from each other in various circumstances. In these ways we elucidate the pre-scientific understanding of distance and direction. If, however, we wish to measure distance and direction to a high degree of approximation or in regions very different in character from that near the surface of the Earth we have to make our tests more precise. We must do so in such a way that our measurements yield the simplest possible scientific laws. It is not always clear, however, when one set of scientific laws is simpler than another set, or when the requirement of ensuring simple scientific laws is being applied in such a way that 'distance' and 'direction' are no longer refinements of our pre-scientific understanding of distance and direction. In such cases and only in such cases are arbitrary decisions needed to make precise our concepts of distance and direction.

How does this account which I have given of distance and direction compare with traditional accounts given by philosophers of science? It is agreed by all writers that one cannot determine the distance between two points (and hence the geometry of Space) until one has determined a congruence standard, that is a standard for judging that two lengths are the same, for example, in terms of rods or light signals. Many writers have, however, urged that it is an arbitrary matter which congruence standard one adopts. Thus Reichenbach: 'A decision has to be made for a definition of congruence. Although we must do so, we should never forget that we deal with an arbitrary decision that is neither true nor false' ([1], p. 19). This seems plainly mistaken, as I have urged earlier. One cannot decide what ordinary words mean. Words in common use mean what they do, and there are recognised tests for determining their applicability, which give them their meaning. There are recognised tests for congruence. If the philosopher is attempting to elucidate the meaning of the concepts we now have, he must describe the criteria which we now use for the application of terms.

Grunbaum, while admitting that one congruence standard is the 'customary' one, urges that there is still a sense in which the decision to adopt it is an arbitrary one.[1] There are two kinds of rule for the use of 'congruent'. First, there are the formal rules which relate 'congruent' to terms like 'equal', 'point', 'line', 'interval' etc. By these rules two intervals are 'congruent' if the 'distances' between their end-points are 'equal', and something is a 'distance' if it conforms to the mathematical axioms for distance which state the logical relations which must hold between different 'distances'. (For these see [5] p. 488.) Secondly there are the material rules which state how to ascertain if two rods are congruent. For Grunbaum the formal rules alone determine the meaning of 'congruent'. How to ascertain whether two physical intervals are congruent is a matter for further choice, and there are many different possible ways of so doing (apart from using the customary tests). Now as 'congruent' is used in pure geometry (see p. 98), no doubt the formal rules suffice to determine meaning. But as a term used in talk about physical shapes and physical distances surely, contrary to Grunbaum, its meaning depends on rules of both kinds. For after all in general, the meaning of any term is given not merely by its logical relations to other terms but by the conditions under which it is proper or improper to apply the term. The meaning of 'green' is given not merely by the fact that it is a colour, different from red, etc., but by the fact that it is the colour of grass, leaves, and gooseberries. So also for 'congruent'. Its meaning is determined not merely by the fact that it is a spatial equality predicate, but by such considerations as that a wooden ruler is approximately congruent with itself when moved across my room.

Grunbaum holds that whatever space is like, there are alternative ways of determining congruence and so of measuring distance. However he holds that it might be that the arrangement of the points of space yielded a way of measuring distance which did not involve using a standard brought from without, such as a rod. For example in a discrete space one could measure the distance between two points by counting the number of points between them. (In a dense or continuous space, this

[1] [2] ch. 1, section D. Grunbaum argues this case against Eddington, who makes my simple point.

would be of no use because all intervals would contain the same number of points. See p. 26 for discrete, dense, and continuous spaces.) Grunbaum calls such a way of measuring distance giving to space an intrinsic metric; and he contrasts it with measuring by such means as rods, which he calls giving to space an extrinsic metric. He claims that if we can give to space an intrinsic metric, so that we measure distance by counting points, we ought to do so because 'the object of a metric . . . is . . . to tell the intrinsic story in so far as possible.' ([5] p. 576.) However, to measure distance by counting points, would involve making the assumption that each point took up an equal volume, and this would seem an arbitrary assumption unless by our normal standards of measurement (i.e. those involved in the extrinsic one) they did so. Grunbaum claims that extrinsic metrics are 'convention-laden', whereas I have suggested that it is part of the meaning of 'distance' that by and large it is the kind of thing which is determined by rigid rods. (For further detailed criticism of Grunbaum see [6].)

True, as I have admitted, criteria in this region may not always be completely clear. There will be obvious cases where the criteria yield one unambiguous result. The distance from a certain point on one wall of my sitting-room to a certain other point on another wall is twenty feet. How to show this is unambiguously clear. Yet what value we give to a distance in the neighbourhood of a very massive body may depend on which of two possible ways we correct our measurements, and the ordinary ways of correcting measurements extrapolated to this case may allow either way of correcting measurements. But although there is a certain vagueness about the correct application of the concept of distance, as of all concepts, in extraordinary circumstances, to the difficulties of which I have drawn attention, there are perfectly clear tests for distance in ordinary circumstances. If the philosopher allows us to introduce completely different tests for 'distance', as Reichenbach and Grunbaum do, he is analysing a concept of his own invention.

In analysing the meaning of 'distance' and 'direction' I have been concerned to describe the primary tests for distance and direction. We do not often make measurements by the methods which I have described. Even on a small scale we do not normally find the straight line between two objects by doing a complicated series of manoeuvres with a ruler, trying to fit it the least possible number of times between the two objects, let alone by finding a common edge to two surfaces shown to be planes by the three-surface test. Rather, we stretch a string between the objects and measure along it. In the next chapter I propose to show that other tests for distance and direction in normal use are secondary tests, and also to describe in a little detail the kind of secondary tests which we use on the cosmological scale. The latter will be a useful preliminary to subsequent analysis of the conclusions of cosmology about the size and geometry of the Universe.

BIBLIOGRAPHY

[1] H. Reichenbach, *The Philosophy of Space and Time* (originally published 1928) (trans. M. Reichenbach and J. Freund), New York, 1957, ch. 1.
[2] A. Grunbaum, *Philosophical Problems of Space and Time*, London, 1964; or second enlarged edition, Dordrecht, Holland, 1973; part i.
[3] Hilary Putnam, 'An Examination of Grunbaum's Philosophy of Geometry' in *Philosophy of Science, The Delaware Seminar* (ed. B. Baumrin), ii, London, 1963.
[4] Arthur Fine, 'Discussion: Physical Geometry and Physical Laws', *Philosophy of Science*, 1964, 31, 156–62.
[5] A. Grunbaum, 'Space, Time, and Falsifiability, Part I', *Philosophy of Science*, 1970, 37, 469–588; reprinted as ch. 16 of the 2nd edn of [2].
[6] Graham Nerlich, *The Shape of Space*, Cambridge, 1976, ch. 8.
[7] A. Einstein, 'The Foundation of the General Theory of Relativity' (originally published 1916) in A. Einstein *et al.*, *The Principle of Relativity* (trans. W. Perrett and G. B. Jeffery) London, 1923.

5 Distance and Direction — (ii) Secondary Tests

Let us now consider some of the tests which we use in practice to measure distance and direction near the surface of the Earth, to substantiate the claim that they are secondary tests. In the examples I shall confine myself to cases of distance being measured from one material object A to another one B in the frame in which A is at rest, viz. the 'proper distance' from A to B.

One test frequently used over short distances is the taut string test for a straight line. Why do we consider that a taut string marks the straight line between two points? Clearly not all stretched bodies will do so. Suppose I stretch a sprig of chestnut tree from A to B. As I stretch it, it takes up a different shape. After I have stretched it to some extent, it does not change its shape any more under further tension. So we may say that it is taut. But it is unlikely to mark a straight line. So why do we say of stretched bodies that the string does and the chestnut sprig does not mark a straight line? Because if we tried to use the smallest possible number of rigid bodies to cover the distance between A and B , they would lie along the line of the string and not along the line of the chestnut spring. Hence the taut string test is only a secondary test for straight line.

Another example of a test in common use to measure distance and direction near the surface of the Earth is the light-ray test. This method assumes that light (and other electro-magnetic radiation — e.g. radio waves) travels at a constant (two-way) velocity along straight lines (i.e. a constant average velocity for journeys to and from a point). If I at O send a light signal to A and it returns after two seconds, and to B and it returns after four seconds (reflections only occurring at A and B respectively) then paths OA and OB taken by light are held to be straight lines, and OB to be twice the length of OA. Now this seems quite obviously not a primary test for distance, for the simple reason that its results are universally recognised as sometimes incorrect. Light does not always travel in straight lines — it does not do so through media of varying density, nor when its path passes very close to material bodies. If what one meant by a straight line was the path taken by light, Grimaldi could (for reasons of logic) not have discovered diffraction. What Grimaldi discovered was that the shadow cast on a surface by an object in the path of a light beam was not the exact geometrical projection of that object, and hence that light is diffracted from a rectilinear path by bodies close to its path. If it is a logically necessary truth that light travels in straight lines, the statement of Grimaldi's discovery would be self-contradictory. Nor of course does light always have a constant velocity — its velocity depends on the medium through which it is travelling. And that light has a constant velocity *in vacuo* is open to experimental refutation. One might, for example, show that it took four times as long for light to travel some distance *OB* and return as to travel *OA* and return, where *OAB* is a straight line in a vacuum on which by the standard of the last chapter *OA = AB*. The fact that this does not happen is experimental evidence that light has *in vacuo* a constant two-way rectilinear velocity, when it is not diffracted. One cannot provide experimental evidence for a definitional truth. Hence the constant rectilinear velocity *in vacuo* of light is not a definitional truth, and so light tests are secondary tests for distance and direction. Hence if the 'light-distance' of an object differs from its distance the former is not the true distance.

In ordinary practice to determine distance and direction we also use a whole group of tests which assume that Euclidean geometry is the true system of physical geometry. Thus if we know the length of a base of a triangle b and the size of the angles α and γ at each end of the base line (see Figure 3) we can calculate the lengths of the other sides by the formula of Euclidean geometry $\dfrac{a}{\sin \alpha} = \dfrac{b}{\sin \beta} = \dfrac{c}{\sin \gamma}$ where $\beta = 180 - \alpha - \gamma$. For measurements over small distances near the surface of the Earth, we are clearly justified in making this assumption. But the question is — do we use such Euclidean tests because we assume that they would give the same results as the rod tests which I described earlier, or do we mean by the distance of an object a result produced by use of certain tests which assume the truth of Euclidean geometry? If the former, then the Euclidean tests are secondary tests for distance. If the latter, then they are primary tests. Now the requirement that the geometry of Space be Euclidean would not uniquely determine the

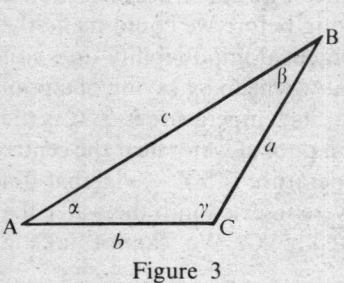

Figure 3

distance of some distant object, and so would have to be employed with other criteria. But still the results given by the tests described in the last chapter might differ from those given by some other method using less rigorous rod tests but assuming a Euclidean geometry, and one could thus define a rod-distance and a Euclidean-distance of some distant object. There would be no logical necessity that the two methods yield the same result, for the first method might show, as we shall indicate more fully in Chapter 6, that the geometry of Space was non-Euclidean. The question is, what would we say if the two methods did yield divergent results — that the first method was the true method and so Space was non-Euclidean, or that the concept of distance was not readily applicable in this case? I shall urge later after I have attempted to describe two non-Euclidean Spaces that it is not a logically necessary truth that Space is Euclidean. It follows from this conclusion that the second answer is false and so that methods that assume a Euclidean geometry are secondary tests, for, if they were primary tests, it would be a necessary truth that Space was Euclidean. For fuller argument on this I ask the reader to wait until Chapter 6.

Between objects further apart than those near the surface of the Earth distance and direction are measured for the most part by methods other than those which I have so far described.[1] I shall shortly run through some of the most important of these, to see the grounds for assertions about the distances of distant galaxies. My claim about these methods is that they too are secondary tests. This is to say, we use them because we believe that the results attained by using them are the same as the results which would be attained by using the primary tests. On the cosmic scale it would be practically impossible for us ever to use the primary tests. There is no surface stretching from the Earth to a distant galaxy on which a straight line could be marked, nor can we construct one. Even if there was such a surface we should die long before we could make the measurements along it. But such mere practical impossibility does not normally have an effect on meaning. What we mean by saying of a pool of water near the surface of the Earth that its temperature is 5° C is the same as what we would mean of saying of a pool of water near the centre of Jupiter if there were one, that its temperature is 5°C — viz. that if a thermometer which passed certain tests were inserted into the water the column would reach the height marked 5° C. We cannot take a

---

[1] The same is true of distance and direction on the very small scale, but as stated in the Introduction, this work would become too bulky were it to provide adequate treatment of problems of the very small scale.

thermometer near the centre of Jupiter to test the latter assertion; but the fact remains that what the assertion means is that, if we could take such a thermometer there, it would record a certain reading. It is often practically impossible to test the truth of statements by using primary tests. But this practical impossibility does not affect the meaning of a statement. My claim is that the methods which we use for measuring directions and large distances in astronomy we use because we believe that the results attained by using them are the same as the results which would be attained by using the primary tests, if we could use these. We must now investigate the methods which we use in practice on the large scale.

In measuring distance and direction on a large scale we assume that relations between bodies found locally hold outside the range for which they have been tested. Clearly in general we are justified in doing so. It is characteristic of science to work on the assumption that the unknown resembles the known, that characteristics of the behaviour of matter found locally hold beyond the ranges for which they have been tested. The kind of circumstances under which we are not justified in extrapolating properties will be considered on pp. 93 ff.

The normal method for measuring the direction of a distant body is from the angle made by light rays emanating from it. The body is stated to be (or rather, to have been at the instant at which the light rays left it) in the direction in which arrive light rays emanating from it. The use of this method assumes that light travels in straight lines *in vacuo* (except when near massive bodies) and that the space between Earth and the stars is very nearly a vacuum. The first assumption has been tested on Earth. The second assumption is justified by considerations of simplicity — the planets obey Newton's laws only if we assume that the space in which they move is very nearly a vacuum. Unless we make this assumption the explanation of planetary motion would become very complicated. Hence we may say that the assumption has been tested within the solar system. We now assume that these assumptions hold for Space far beyond the solar system, and hence can estimate the direction of distant bodies.

The most important method for measuring the distance of near-by stars is the method of parallax and is a more complicated form of the kind of method described on p. 85. This assumes, as well as that directions can be estimated in the way just described, that the geometry of Space is Euclidean. This has been found to be so on Earth, and it is now assumed that it holds for Space far beyond the Earth.

The method of parallax uses a short distance measured by other methods, that is ultimately by rigid body or taut string methods, this distance being the base line. The angles at each end of the base line made with the base line by light rays impinging from the distant object are measured and the distance of the object calculated by the formulae on p. 85. This, basically, is the method by which the distance from the Earth to the Sun and the Moon and the planets was established (the grounds for using light rays to ascertain direction being, when these distances were originally established, weaker than those stated above). The first stage was to ascertain the radius of the Earth. This was done in the fourth century B.C. by measuring the different angles subtended at different places on the Earth by light rays from the Sun. But it could be done by rigid body methods, if men were prepared to take the necessary trouble. The circumference of the Earth could be measured and the radius calculated from it by the use of the formula of Euclidean geometry that the circumference of a sphere is 2π times its radius. That the Earth is approximately a sphere can be shown by measuring many great circles on it (that is, lines on the surface of the Earth, marking the shortest distance along the surface between any two points on the line) and showing that these are of approximately equal length. We then obtain a base line for our astronomical measurements by measuring the distance along the surface of the Earth between two places on it A and B and then from the known radius of the Earth calculating, using Euclidean geometry, the distance along the straight line through the Earth between A and B. This gives the length of the base line b. The use of this method informs us where the distant object was when the light left it. (For a base line long relative to the distances being measured, we have to add a correction to allow for the different periods of time taken by the light to come to the two points on the base line from the distant object. For a base line like one on the surface of the Earth, this factor may be ignored.) We can thus measure the distance from the Earth to the Sun, Moon, and planets.

We can then infer, as Kepler did, from the change of the Sun's position relative to the Earth the path of the ellipse traced out annually by the Earth round the Sun. The Earth moves some 184 million miles from one side of the Sun to the other during the half-year. We can now use a straight line from one side of the ellipse to the other as base line for the calculation of much larger distances by the method of parallax. This is normally done as follows.

We measure the angles of a star at different instants during the year and find two equal and opposite bearings of the star ($\alpha = \gamma$ in Figure 4). The Earth's path being nearly circular, the base line b is at all periods of the year approximately 184 million miles. The angle $\beta/2$ ($90° - \alpha$) is then known as the annual parallax of the star. $a = c = \dfrac{b}{2\cos\alpha}$. The first star

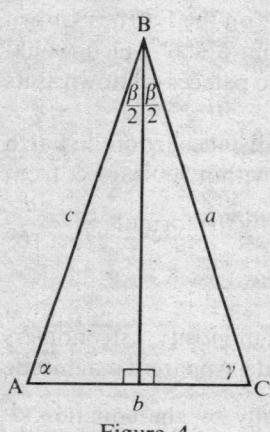

Figure 4

parallax was calculated in 1838 by F. W. Bessel, who found that the annual parallax of the star 61 Cygni was 0.31″, and hence that the star was at a distance of some 54 million million miles from the Earth. Astronomical distances are normally measured in parsecs. A star with an annual parallax of 1″ is said to be at a distance of 1 parsec from the Earth. Hence 61 Cygni lies at a distance of some three parsecs. The nearest star to the Earth, Proxima Centauri, lies at a distance of some one-and-a-half parsecs.

The calculation of distance by parallax thus works on the assumption that certain locally tested regularities (Euclidean geometry, the rectilinear propagation of light) hold far outside the range for which they have been tested. Thereby the range within which distances can be measured is enormously extended. Using the Earth's orbit as base line astronomers can calculate distances up to 100 parsecs. A more complicated method known as the method of statistical parallax is used to calculate distances up to 500 parsecs. I shall not describe this method as it is somewhat complicated (for a brief description see [1] pp. 29–34). Hence I shall state without proof my thesis about it that it works by assuming that certain regularities observed to hold locally hold far outside the range for which they have been tested.

A variety of further methods are used to calculate greater distances, the most important for medium distances being the 'headlights' method. Certain stars, especially the Cepheids, are variable stars whose luminosity increases and decreases in a characteristic way, the period of oscillation being constant for each Cepheid but varying for different Cepheids from one day to many years. The distance of some of these stars near to us can be calculated by the method of statistical parallax. We assume that the locally tested law that the luminosity impinging on an observer from a distant object varies inversely as the square of the distance of the observer from it holds far outside the range for which it has been tested (a correction has to be added when the distant object is in motion relative to the observer. We have evidence from the absence of significant Doppler shift—see p. 91—that such corrections would be insignificant for stars of our own or near galaxies). Hence, knowing the distance from ourselves of near-by Cepheids we can calculate how bright they would appear to an observer at a distance of ten parsecs. The luminosity impinging from a star on an observer on the Earth is known as its apparent luminosity. The luminosity from a star which would impinge on an observer at a distance from it of ten parsecs is known as its absolute luminosity.

From its apparent luminosity (l) and known distance from the Earth in parsecs (d) we can thus calculate for any star within 500 parsecs from the Earth the absolute luminosity (L) of the star by the formula $\dfrac{L}{l} = \dfrac{d^2}{10^2}$

or $L = \dfrac{ld^2}{100}$.

Instead of dealing in apparent and absolute luminosity, astronomers normally deal in apparent and absolute magnitude. Apparent magnitude is related to apparent luminosity logarithmically by the equation $\dfrac{l_1}{l_2}$ $= 100^{(m_2 - m_1)/5}$, where l_1 and l_2 are the apparent luminosities of two stars and m_1 and m_2 their apparent magnitudes, the apparent magnitude of certain stars being fixed by ancient convention. Absolute magnitude is the apparent magnitude which would be calculated from luminosity impinging on an observer at a distance of 10 parsecs. Absolute magnitude (M), apparent magnitude (m) and distance in parsecs $\left(\dfrac{1}{w}\text{ where } w \text{ is the star's parallax}\right)$ are therefore related by the formula $M = m + 5 + 5 \log w$.

We can by these means calculate the absolute magnitude of Cepheids within 500 parsecs at different instants during their luminosity cycle. Astronomers found a regular linear relationship between the logarithm of its period and the mean absolute magnitude of a Cepheid. Hence if a Cepheid outside the range of calculation of distance by statistical parallax behaves like those within the range, then we can infer by this formula from its observed period to its mean absolute magnitude and hence its mean absolute luminosity. We observe its mean apparent luminosity. Given again that $L = \dfrac{ld^2}{100}$ outside the range for which this law has been tested, from mean absolute luminosity (L) and mean apparent luminosity (l) we can infer the distance of the Cepheid.

From the distance of a Cepheid we can calculate the approximate distance of near-by stars. A star is close to others if it is tangentially close and radially close. Given the locally tested law that light travels through space in straight lines, stars close to others on a photograph of the sky will be in fact tangentially close to them. The evidence that a tangentially close star is radially close to another is that it is of similar apparent magnitude and has a similar velocity of approach or recession. For we know from observations within the region where distances can be calculated by the method of statistical parallax that tangentially close stars of similar apparent magnitude which have a similar velocity of approach or recession are normally also radially close to each other. We assume that this locally tested characteristic of stars also holds outside the range of observation. We ascertain a star's velocity of approach to or recession from the Earth by studying its Doppler shift. We know from observations on the Earth that the spectrum of light from a body B as observed at A is shifted towards the violet end of the spectrum if B is approaching A, and towards the red if B is receding from A. We assume this to hold for stars and galaxies far outside the region for which it has been tested and know of no other explanation of a red-shift (apart from the gravitational red-shift predicted by General Relativity which would not normally be relevant). We also assume the locally tested law that the elements of matter have the same spectra in all places, and so that there are only certain possible emitted spectra. Hence from a received spectrum, given that all lines have been shifted together, we can infer what was the emitted spectrum and so infer velocity of recesion or approach. That such an inference can always be made confirms the law that the elements are the same at all places and have there the same spectra. The proportion by which the wave-lengths of received light exceed those of emitted light due to a velocity of mutual recession of source and observer is known as the Doppler shift of the source.

Using all these tools we can infer which stars are near to a given Cepheid and hence their approximate distance from the Earth. Once we know which stars are near to each other, we realise that the Universe is populated by large groups of stars, called galaxies, to one of which our solar system belongs; and that the galaxies keep together in larger units called clusters. We can infer the distances of near-by galaxies from the Cepheids in them.

There are many similar methods of calculating distances of stars in our own or near-by galaxies which lie outside the range of calculation by statistical parallax. They all depend for their operation, like the two methods so far described, on assuming some locally tested physical regularity to hold outside the region for which its operation has been tested.

Having by such means found the distance of near-by galaxies, we can calculate from their apparent magnitude and diameter, their absolute magnitude and actual diameter. We then notice that the absolute magnitude and the diameter of galaxies varies only within a narrow range. Assuming once again that what holds locally holds more widely, we are now in a position to judge the distance of a distant galaxy from its apparent magnitude. For we know within a range its absolute magnitude and given the inverse square law of diminution of luminosity can infer thence its distance. (Evidence from experiments on Earth shows that a correction has to be added to the formula for a light source with a velocity of recession from the observer, so that a source with a velocity yielding a Doppler shift of z lies at a distance $\dfrac{1}{1+z}$ times that which would otherwise have been estimated.) A similar inference can also be made from the apparent diameter of a galaxy indicated by the angle which it subtends for an observer on Earth (see Figure 5). If we know within a range the actual diameter of a galaxy, we can then infer, within a range, its distance from the Earth, given the rectilinear propagation of light and Euclidean geometry. This method as stated is not of much use for other than near-by galaxies since the angle subtended by others is too small to be determined accurately. But a similar method can be applied to clusters of galaxies. We observe that galaxies of similar magnitudes with similar recessional velocities appear close together on photographic plates. We infer that they form a cluster of galaxies close to each other.

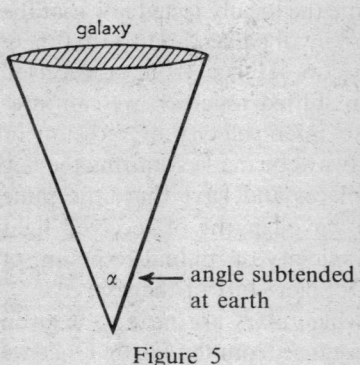

galaxy

α ← angle subtended at earth

Figure 5

We find the actual diameters of near-by clusters, infer that what holds locally holds outside the tested range, and thus infer from the angle subtended by them at the Earth the distances of clusters of galaxies. Distances of individual galaxies or clusters calculated by these methods are obviously very rough, but the methods are very useful for calculating the average distance of large numbers of galaxies or clusters, when errors made in calculating the distance of individual galaxies cancel out.

The most important method for ascertaining the distance of the most distant galaxies is from their red-shift. Assuming on the grounds stated earlier that red-shift indicates recession, we note a functional relation between the distance of a galaxy and the velocity of its recession indicated by its red-shift. The astronomer E. P. Hubble who first noted this relation claimed that this was a linear relationship between velocity (V) and distance (d) $V = Hd$ (H being 'Hubble's constant'), but some later astronomers have claimed that the relation is non-linear. But granted that observations on galaxies whose distances can be ascertained by such other means as the method of apparent magnitude supports Hubble's law, we can go on to use it to establish distances of the most distant galaxies observed, i.e. distances up to 1,000 m. parsecs from ourselves. (Although there are the good grounds described earlier for holding that red-shift shows velocity of recession in the case of stars and galaxies, there are grounds for suspecting that the red-shift of quasars may have a different cause. Quasars are strange star-like objects first discovered in the mid-sixties about whose nature scientists are still very puzzled.)

So by continual extrapolations from local observations, the distances of more and more distant objects have been ascertained. I have sketched only in the briefest outline a few of the methods used. I have given the outline to illustrate the methods of finding out the distance and direction of distant objects by further and further generalising from local patterns of phonemena. Since we use these methods because we believe that certain patterns of phenomena hold beyond the range for which they have been observed to hold, if we find reason for no longer holding this belief about certain patterns of phenomena, we ought to abandon any methods of estimating distance and direction which presuppose its truth.

There could be two distinct kinds of reason for believing that certain patterns of phenomena do not hold beyond a region for which they had originally been observed to hold. The first kind of reason would be that subsequent observations did not confirm the original observations that one property in question held in a local region. Thus using some method I we may have noted that the Universe within the range of study by that method had a property P. Assuming that P holds outside that range of observation, we can then use a method II which embodies this assumption to find further distances. Suppose now that by more careful, precise and painstaking use of method I, we find that P does not in fact hold of the Universe within the range of observation by method I and hence there are no grounds for assuming it to hold outside that range. In that case the use of method II to ascertain distance would not be justified. Astronomical methods of ascertaining distance are continually being modified in the light of such considerations. Thus all methods which work from the luminosity of distant objects need an estimate of the amount of light absorbed by interstellar and intergalactic dust. Given absolute magnitude and apparent magnitude, we can only infer distance if we judge insignificant or give a value to this factor. Initially the astronomer may judge this factor to be insignificant, and by some method II (e.g. inferring the distance of a galaxy from its apparent magnitude as described on p. 92) may infer its distance. However, more careful estimates of the amount of intergalactic absorption may lead him to revise his method II.

The second kind of reason for believing that certain patterns of phenomena do not hold beyond the region for which they had originally been observed to hold could arise even if subsequent observations confirmed that the property did hold in the region studied by the original method. This kind of reason would be that the assumption that it held beyond the region for which it had been observed to hold proved incompatible with the reasonable supposition that various other properties held beyond the range for which they had been observed to hold. Thus suppose that by using some method I, we observe that properties P, Q, R, and S hold of the region of the Universe which can be studied by that method. The assumption that P holds universally means that we can use a method II for ascertaining further distances. The assumption that Q, R, and S hold universally means that we can use respectively methods III, IV, and V for ascertaining further distances. Suppose now that the use of method IV gave different estimates of distance from the other methods. Then either R does not hold beyond the range studied by method I, or P, Q, and S do not.

If we have no general theory of physics on a cosmological scale, the simplest assumption to make is that the odd property R does not hold outside the range of observation by method I, and so that method IV is not a justifiable method of ascertaining distance. However, if we have a cosmological theory, the situation is different. On the basis of observations that properties P, Q, R, and S hold locally a scientific theory may be set up which gives for those observations a more simple and coherent explanation than any other theory compatible with them. It may, however, be a consequence of the theory that while R holds universally, P, Q, and S do not hold universally. Thus it may be a consequence of the theory that P, Q, and S only hold in a small region and that in a wider region a property different from P, Q, and S would hold, but that the region studied was too small for a difference from P, Q, and S to be noticed. Hence P, Q, and S would not be extrapolatable properties. The use of methods II, III, and V which assume them to hold beyond the original range would not be justified and would have to be corrected in the light of the theory. What is happening here is that the theory reveals that one or more local properties, in our example R, are more fundamental than others and the assumption that these hold universally will mean that others do not. Cosmological theories developed from the General Theory of Relativity and rivals to it have involved such modifications to methods of estimating distance. The General Theory cast doubt on the general validity of Euclidean geometry. Since all the secondary tests for distance assume that geometry is Euclidean, they must be modified if that assumption turns out to be unjustified. Various forms of General Theory and other cosmological theories lead to different conclusions about the relation which they claim holds universally between the distribution of matter and the geometry of Space. In so far as any one theory is well substantiated, we must assume that the relation which it shows to hold locally holds universally. Hence from our knowledge of the density of matter on a large scale we can infer the geometry of Space on a large scale, and hence ascertain that it is different from one which to a high degree of approximation holds on the small scale, viz. Euclidean geometry. The theory by extrapolating a certain property and assuming it to hold universally showed that a different property would not hold universally. Whether we are to adopt a theory which assumes a property P to hold universally, or a theory which assumes another property R to hold universally, depends on the relative simplicity of the two theories. The character of evidence for any general cosmological theory like General Relativity and the conclusions about the geometry of the Universe which can be derived from it will be examined in Chapter 14. Clearly there are various difficulties in building up such a theory — we shall have to show, for example, how we can reach a conclusion about the density of matter without assuming a certain geometry to hold. But assuming that we can by these means show that a certain geometry, other than the Euclidean which to a high degree of approximation holds locally, holds universally, we should have to modify such methods of calculating distance as the method of parallax and the method of apparent magnitude which take for granted the universal applicability of Euclidean geometry.

So in these ways the methods described of ascertaining distance and direction can be and are being continually refined in the light of experience. The fact that all the methods described of ascertaining distance and direction beyond the Earth are open to review and refinement indicates that the methods are secondary tests. We use secondary tests because we believe them correlated with primary tests. To the extent that we have evidence that this correlation does not hold we amend the secondary tests. If empirical evidence could show that a test for distance did not give good results, that shows that the test is a secondary test. Empirical evidence could not show this for primary tests, since empirical evidence cannot refute a definition.

It is a consequence of these points that various concepts used by cosmologists related to the concept of distance are not in fact concepts of distance. Cosmologists have defined several different 'distances' of objects differing in respect of the method of estimating them. Thus there are 'distance by apparent size', 'luminosity distance', and 'parallax distance'. The distance by apparent size of a galaxy G is $\dfrac{D}{2 \tan \alpha}$, where D is the average diameter of a local galaxy and α the angle subtended at the Earth by G. The luminosity distance of a galaxy G is $10 \sqrt{\dfrac{l_1}{l_2} \dfrac{1}{(1+z)}}$ parsecs, where l_2 is the apparent luminosity of G, l_1 the absolute luminosity of an average local galaxy, and z is the Doppler shift of the galaxy. Parallax distance is $\dfrac{b}{2 \cos a}$, where a is the angle between b, the diameter of the Earth's path round the Sun and light rays impinging on the Earth from the distant object, measured at points on the Earth's path where the angles between light rays from the distant object and the diameter are equal and opposite.

Now we would only use these simple formulae for calculating the distance of a distant object on certain assumptions — that the geometry of Space is Euclidean, that light travels in straight lines, that there is virtually no absorption of light by interstellar dust, etc., etc. Some of these assumptions are probably justified and some are probably not. Since cosmologists do not alter their estimate of the luminosity distance of a galaxy when provided with information about the amount of light absorbed by intergalactic dust, it cannot be an estimate of the distance of a galaxy. Similar considerations apply to the other distances. It may be easier to construct a coherent cosmological theory if we deal with luminosity distance, etc. instead of distance, but luminosity distance, etc. are not distances.[1]

---

[1] Hence McVittie is wrong to claim that 'Different procedures may, and in general do, give rise to different distances and there is no one of them which can be labelled as the "correct" distance' ([3] p. 147). Different procedures give different measurements, but they are not all measurements of distance. Trautman may be right when he claims that 'proper distance' (viz. distance relative to the frame in which the first object is stationary) 'is of no practical use', in cosmology. But he is misguided to recommend the use in cosmology of

BIBLIOGRAPHY

For accounts of the methods used by astronomers for measuring distance see:
[1] G. C. McVittie *Fact and Theory in Cosmology*, London, 1961.
[2] M. Berry, *Principles of Cosmology and Gravitation*, Cambridge, 1976.
[3] G. C. McVittie, *General Relativity and Cosmology*, London, 1956, ch. 8.
[4] H. Spencer Jones, *General Astronomy*, 4th edn, London, 1961.

For philosophical commentary see:
[5] J. D. North, *The Measure of the Universe*, Oxford, 1965, ch. 15.[1]

---

luminosity distance on the grounds that 'we seek a definition of an observational nature, which can actually be used by astronomers to determine the distances of distant nebulae.' For luminosity distance will only measure 'the distance of distant nebulae' if it as corrected, as appropriate, in the ways indicated in the text. See A. Trautman's lecture in A. Trautman, F. A. E. Pirani and H. Bondi, *Lectures on General Relativity* (Englewood Cliffs, New Jersey, 1965) p. 242.

[1] On pp. 347 f. of this work Dr North has attempted to show that it is a consequence of the Robertson–Walker metric (the metric — see my Chapter 11 — applicable to a homogeneous Universe), that for three 'fundamental particles', that is roughly, clusters of galaxies, P, Q, and R not on a spatial geodesic (viz. not on a straight line) it is possible that we might find, where k (see Chapter 11) = $+1$ or -1, $d(PR) > d(PQ) + d(QR)$, where $d(AB)$ is the proper distance from a fundamental particle A to another one B.

It might appear that if North's arguments were correct, he would have shown that the concept of proper distance had no application on a cosmic scale, for is not $d(PR)$ by definition the shortest path from P to R? So if $d(PQ) + d(QR)$ were less than $d(PR)$, would we not have found a shorter path from P to R? This would be so only if P and Q were at rest relative to each other. For $d(PR)$ being a proper distance must (see p. 67) be measured along a line of constant length stationary relative to P, and the line from P through Q to R would only be such a line if Q was stationary relative to P. North does not purport to show that his proof holds for this particular case.

6 The Geometry of Space

Let us begin discussion of the question of the geometry of Space by making the well known distinction between pure and physical geometry. The basic element of pure geometry is the point, a term usually left undefined. Which other terms are undefined will depend on the way in which the geometry is axiomatised; but 'distance' and the relation of being 'collinear with' are often also undefined. Other terms such as 'line' and 'plane' may be defined in terms of the undefined terms, axioms set up and theorems proved therefrom. We can then give any interpretation we like to this axiomatic system, and if the axioms are true, their consequences, the theorems, will be true also. Different sets of axioms or different interpretations of the axioms lead to different 'spaces' in the metaphorical sense of the term. Thus a Hausdorff space means a collection of 'points', the relations between which satisfy Hausdorff's axioms, while 'momentum space' is a collection of 'points', each 'point' representing a different possible state of momentum of a particle.

Physical geometry interprets the terms of physical Space in their normal sense. Thus a solid, in the geometer's sense, is any place, any volume of space which is or could be occupied by a material object. (The geometer's 'solid' is to be distinguished from the physicist's 'solid'. For the physicist a solid is a material object which does not change its shape readily when subject to deforming influences.) Any solid is fully enclosed by a surface; and anything which encloses or could enclose a material object, or is part of something which could, is a surface. We can point to examples of 'surfaces' to clarify more fully the meaning of the term — thus the top of my desk is a surface, and so is the side of my house.

Analogously any proper part of a surface which encloses a material object will itself have a boundary, which is a line; and a line is anything which forms such a boundary or could do so, or is part of something which could. Again, examples of 'lines' will bring out more fully the meaning of the term — the edge of the top of my desk is a line, and so is the western edge of the southern wall of my house. Any boundary to a line is a point, and a point is anything, which could form such a boundary. A point is (see p. 26) the smallest bit of space there could be.

Given that space is not discrete (see p. 26), then a point will have no volume at all. Euclid [1] defined a point as 'that which has no part'. But in whatever sense space is discrete, then in that sense its smallest bits would have a volume. Given that space is not discrete, anything very small in comparison with the distances in which we are interested, will serve for practical purposes as a point — e.g. a dot on a blackboard. Analogously, Euclid's defined a 'line' as 'a length without a breadth' and a 'surface' as 'that which has breadth and length only'. Even if space is not discrete, for practical purposes anything may be regarded as a surface if its height is negligible in comparison with its breadth and length and other distances of interest, and as a line if its height and breadth are negligible in comparison with its length and other distances of interest. If however space is discrete in some sense, then in that sense its lines and surfaces will have very small volumes. The tests for a line being a straight line, and a surface being a plane I have described in Chapter 4.

The task of the physical geometer is to choose such axioms that true theorems about points, lines, planes, etc. always result. The most precise and best-known form of geometry is metrical geometry. To set forward the metrical geometry or metric of Space is to say what relations of distance and direction between points entail other such relations and also propositions about areas and volumes. Thus a metrical geometry may tell us that, if we draw a line, each point on which is equidistant from a certain point, to be called the centre (viz. we draw a circle), then a straight line which touches this line without cutting it (a tangent to the circle) will make a right-angle with the straight line drawn from the point of contact to the centre. Again it may tell us if we draw straight lines of a certain length from A to B and from B to C and the angle ABC is a right-angle what the length of AC will be.

The quantity of a surface is measured by its area, of a solid by its volume. These are derivative quantities calculated from distances (e.g. distances along the edges of the surface) by formulae of the metrical geometry, which vary with the geometrical system. Thus in Euclidean geometry the area of a circle of radius r is πr^2, r^2 being the area of a square of side r. However, in order that the 'area' and 'volume' of the geometry be the area and volume about which we ordinarily talk, the formulae must satisfy the following requirements. Two figures have the same area if they are congruent two-dimensional entities. Two figures have the same volume if they are congruent three-dimensional entities. A figure A has $\frac{m}{n}$ times the area (or volume) of another figure B if, for some k, A can be divided into km areas (or volumes) and B can be divided into *kn* areas (or volumes) all of which are congruent with each other. Two figures are congruent if they have the same shape and distances between corresponding points on each are the same. Thus a square of side two inches is four times the area of a square of side one inch because it can be divided into four squares of side one inch.

The most celebrated metrical geometry is of course that of Euclid. In the third century B.C. Euclid formalised current knowledge about physical geometry into an axiomatic system. The modern pure metrical geometer takes over the axioms and definitions of Euclid's geometry and, leaving the terms uninterpreted, modifies the former in various ways and investigates what theorems result from what modifications.

A more general form of geometry is topology. The topological properties of a body are those which are invariant under continuous deformation and include such properties as neighbourhood (a point being next to some other point), and enclosure (a point lying within some enclosing set of points). However you push and pull a body, so long as you do not cut it or join points previously separate, it retains its topology.[1] Thus, to take two-dimensional examples, the square and the circle have the same topology, for you can make a circle out of a square by pushing in its four corners. But a circle and a two-dimensional letter 'B' do not have the same topology. For you would have to cut a letter 'B' to make a circle out of it; and join parts of a circle previously separate to get a letter 'B'. To take three-dimensional examples, a cube and a sphere have the same topology, but a cube and a torus (a solid ring) have different topologies.

One can derive theorems about topology either by assuming a system of metrical geometry or by constructing an independent topological axiom system. If a body is stated to have certain topological properties, then, since it will continue to have those properties under continous deformation, we can suppose it to have any shape we like consistent with those properties and use metrical geometry to infer other topological propositions about it. Alternatively the topologist can set up an independent system and construct a set of axioms about the separations, neighbourhood relations, and enclosures of points, from which he can derive theorems about these and other topological properties. As with metrical geometries, there can be pure or physical topologies. The pure topologist leaves his terms uninterpreted. The physical topologist

---

[1] Strictly speaking, you are allowed to make a cut so long as you join the body up again in exactly the same way as it was previously joined.

understands them in their ordinary physical senses, some of which I have earlier described.

Having made the distinction between pure and physical geometry, we shall henceforward normally be concerned only with physical geometry.

Now what is the true physical metrical geometry of Space? Which propositions about particular distances and directions entail which other propositions about particular distances and directions (and propositions about areas and volumes)?

The geometer is concerned with the straight lines, planes, etc. which join points. Hence, as we saw in Chapter 4, in order that his terms be understood unambiguously, we have to specify a frame of reference by which places and so points are to be re-identified. In order to state the laws of physics we must, as we saw in Chapter 3, reidentify places by the most basic frames of reference that there are. Hence, in order to describe the character of Space in which bodies governed by the laws of physics stay still or move, we must use such frames. If there was only one most basic frame of reference, the geometry of physical Space would be that estimated by reference to that frame. As, at any rate on the cosmological scale, there are many equibasic frames, then only if the same geometry results by whichever of these frames we make our measurements, can we talk of the geometry of Space. If a different geometry would be measured relative to different equibasic frames then we could not talk of the geometry of Space *simpliciter*, only of the geometry of Space relative to some frame. The evidence of modern cosmology to be given in Chapter 11 is that the same geometry will be measured relative to each equibasic frame. Hence we can talk of the geometry of Space.

Until the middle of the nineteenth century it was universally believed to be a necessary truth and 'necessary' in a stronger sense than 'physically necessary', that the geometry of Space was Euclidean. Classically, Kant wrote of 'The apodeictic certainty of all geometrical propositions' by which he meant that the applicability to any physical Space of the theorems of Euclidean geometry was an *a priori* truth.[1] He argued that these theorems were the only ones of which it made sense to suppose that they applied to physical Space. It was not conceivable that a different geometry might apply to physical Space. Yet it seems in no way obvious that some of the theorems of Euclidean geometry are not mere logically contingent truths. If some theorems of Euclidean

---

[1] [5], A24. Kant held geometrical propositions to be a synthetic *a priori* truths. But I shall subsequently represent him, as explained on p. 4, as claiming that they are logically necessary truths.

geometry are mere contingent truths, then some of the axioms must also be contingent truths. The axiom which always seemed to geometers less obvious than the others, was Euclid's famous fifth postulate which can be stated in a form (given the truth of the other axioms) equivalent to the original, known as Playfair's axiom as follows:

For any line *L* and point *P* not on *L*, there is one and only one line *L'* through *P* parallel to *L*. (Two lines are parallel if they are in the same plane and, however far extended, never meet.)

For two thousand years geometers attempted to derive this postulate from the other axioms but failed and finally proved that the postulate was independent. Hence an alternative geometry could be obtained by altering the postulate. Geometers have subsequently obtained other self-consistent geometries by altering other axioms of Euclid. Some of these geometries, including ones which we shall consider in Chapter 7, may perhaps prove not to be logically possible physical geometries. 'Point' therein cannot be interpreted as very, very small place and 'distance' as distance. They may however be perfectly self-consistent systems of pure geometry and have applications other than those in which the terms are interpreted in their normal physical senses. Yet some of the geometries do seem to describe genuine alternatives to Euclidean geometry which might hold of physical space.

The two of these differing least from Euclidean geometry are the two original non-Euclidean geometries developed by adopting a different parallel axiom to Euclid's fifth postulate. These are the hyperbolic geometry invented by Bolyai and Lobachevsky, and the elliptic geometry, invented by Riemann and Klein. Hyperbolic geometry is obtained from Euclidean geometry by substituting for Euclid's fifth postulate, the postulate that:

For any line *L* and point *P* not on *L*, there is more than one line *L'* through *P* parallel to *L*. (If there is more than one parallel line, it can then be proved that there are infinitely many.)

Euclid's other axioms remain. But with this different axiom, many of the theorems of hyperbolic geometry are radically different from those of Euclidean geometry. For instance, the interior angles of a triangle sum to less than 180°. The angle sum is smaller, the greater the length of sides of the triangle. The ratio of the circumference to the diameter of a circle is greater than π. There can be no similar figures which are not congruent. This means that if you take some figure and compare it with another figure with sides of different lengths but having the same ratios to each other — say a pentagon with each of its five sides of equal length compared with another pentagon also having each of its five sides of equal length to each other but longer or shorter than those of the other pentagon, then the angles of the two figures and hence the shape of the figures will be different. There are many other such differences from Euclidean geometry.[1]

Elliptic geometry is obtained by amending the fifth postulate to:

For any line L and any point P not on L, there are no lines through P parallel to L.

Euclid's first three postulates, if interpreted as they normally are, have to be modified to make this new postulate consistent with the rest of the Euclidean system.[2]

The properties of elliptic geometry include the following. The interior angles of a triangle sum to more than 180°. The angle sum is greater, the greater the length of the sides of the triangle. The ratio of a circumference to the diameter of a circle is less than π. There can be no similar figures which are not congruent.

---

[1] For details see any textbook of non-Euclidean geometry, e.g. [2].

[2] Thus Euclid's second postulate, as written by Euclid, states that a finite straight line can be produced continuously in a straight line. This is formally consistent with elliptic geometry. But Euclid's postulates are normally taken in a more stringent sense than the literal in order that his theorems may be derived from them rigorously. The second postulate was taken before the nineteenth century as stating that a straight line can be produced infinitely (viz. without meeting itself again) and later more rigorous formulations of Euclidean geometry made explicit this sense of the postulate. In this form the second postulate is inconsistent with the new fifth postulate and has to be modified back to its original form to produce a consistent system of elliptic geometry.

Riemann originally developed spherical geometry, using this new fifth postulate but making different slight modifications to Euclid's first three postulates as these are normally taken. Klein modified this to elliptic geometry. The distinguishing characteristic of spherical geometry is that to every point there is an anti-point, viz. all straight lines which meet at any point P meet also at a different point P'. If the Universe had a spherical geometry and light travelled in straight lines, then the rays of light leaving a star at a point S would all intersect at another point S'. To an observer using visual observation it would appear that there was a star at S', but there would in fact be at S' a mere optical image. No evidence of the existence of such anti-points having been found, cosmologists prefer elliptic geometry as their geometry of positive curvature (see p. 104 for this term). Spherical and elliptic geometries have, with the exception discussed, virtually the same properties. The term 'Riemannian geometry', it should be noted, is often used of a wide class of geometries including the three non-Euclidean geometries so far discussed, and not simply for the geometry invented by Riemann.

Now consider a sphere centered on a point P found to have a surface area $4\pi r^2$. The radius of the sphere S will be in all three geometries $\int_0^r \dfrac{dr}{\sqrt{1 - Kr^2}}$ where K is some value, positive, negative or zero, called the curvature of space. If K is a constant throughout space, the geometry will be one of the three so far considered, which are therefore known as geometries of constant curvature. If $K = 0$, space is Euclidean; the radius of the sphere is r. If K is negative, the space is hyperbolic; the radius is less than r — in other words, a sphere of radius u has an area of more than $4\pi u^2$. If K is positive, the space is elliptic; the radius is more than r — in other words, a sphere of radius u has an area of less than $4\pi u^2$. K is a measure of the divergence of the geometry from the Euclidean. The larger its positive or negative value, the greater the difference from $4\pi u^2$ of the area of a sphere of radius u, and the more marked are all other differences from Euclidean geometry. (In all cases every point on the surface of a sphere of radius $4\pi r^2$ centred on P will be said to have a 'coordinate-distance' r from P. We shall need this term later.)

Now the non-Euclidean geometries which I have set forward appear to represent genuine logically possible alternatives to Euclidean geometry which might hold of physical Space. Since the nineteenth century this has generally been accepted, and, if it is accepted, it confirms the point which I made in Chapter 5 that it is not a presupposition of measuring distance and direction that the geometry of Space be Euclidean. This being so, I shall not argue in great detail against the Kantian position.[1] For the moment I shall take for granted the falsity of the Kantian position, but when I have discussed in more detail the topology of physical Space I shall be in a position to present one brief conclusive counter-argument to it.

If we assume then that there are alternative metrical geometries to Euclid's, to find out which is instantiated in our Space we would ideally have to measure angles and distances in the ways described in Chapter 4 to find out which relations between figures held in it; or if direct measurement were not possible, we would have to use the indirect methods set forward in Chapter 5.

Physical geometry thus acquires the character of a physical theory. A physical theory is shown false if predictions from it prove on observation to be false. A physical theory is confirmed in so far as its predictions are confirmed by observation and it unifies them in a more simple and coherent way than can any other theory compatible with them. So the evidence that the axioms of a system of physical geometry are true is that the predictions from the system turn out to be true and that the axioms form a more simple and coherent system than any other system compatible with the predictions. Thus, suppose we measure the sum of the interior angles of a number of triangles and find that it is in each case less than 180°, then, if we suppose that we have made our measurements correctly, we have shown that the geometry of Space is not Euclidean. If other predictions of hyperbolic geometry are confirmed and the hyperbolic system is a more simple and coherent way of accounting for these than any other system, then observational evidence has shown that the hyperbolic geometry is the geometry of Space. The larger the region studied and the more accurate the measurements made, the better confirmed will be any conclusion about the metric of Space as a whole. Thus both elliptic and hyperbolic geometries are scarcely distinguishable from Euclidean in small regions; and so the larger the region taken or more accurate the measurements the better the evidence about which of the three geometries is the true physical geometry of Space.

Other systems of geometry can be obtained from Euclid's by modifying other of his axioms. But the Euclidean, elliptic (and spherical, see p. 103, note 2), and hyperbolic geometries have, unlike most metrical geometries, the property of being congruence geometries. A geometry which permits a figure congruent to a given figure to be constructed in every region of space is called a congruence geometry. Another way of describing a congruence geometry is to say that it is characterised by the axiom of free mobility that an ideally rigid body could be moved throughout Space without its shape changing. If the geometry of Space is uniformly of one of the three types referred to and possesses everywhere the same curvature, then an ideally rigid body could be moved throughout Space without it changing its shape. Suppose on the other hand the curvature varied so that it was now zero (viz. the geometry was Euclidean), now of considerable positive curvature (viz. the geometry was elliptic), the axiom of free mobility would not in general hold.

Now the supposition that the geometry of Space is not a congruence geometry is incompatible with the supposition that there can be an ideally rigid body. All measuring instruments would be deformed under transport. However, one can still maintain that the distance between two points on certain measuring rods and the distance between two points on all rods, which continued under transport to coincide with points on them, was not deformed, and then the geometry of Space would be the geometry measured by those points on those rods. We saw in Chapter 4 that this was what Einstein did in 1916. For him the geometry of Space was that measured by very short straight rods of negligible thickness. We concluded that, although this seemed a slightly arbitrary supposition, it could be justified on the grounds of the simplicity of the resulting physical theory, since the deformations which it postulated were only very slight, and so the ordinary understanding of length remained as that which is measured approximately by all rods which preserve among themselves approximate coincidences in a variety of circumstances. Einstein claimed, on his understanding of what constituted a true measuring instrument, that the geometry of Space was of variable curvature. Inside massive bodies it was of positive curvature, immediately outside massive bodies of negative curvature, and far from massive bodies Euclidean. Overall, however, he argued in 1917, it had a positive curvature.

Now Poincaré [6] argued that we could always choose, whatever observations we made, to say that space had any geometry we like simply by postulating that rods had had their length altered in some way. For geometries of different topology, as I shall show later and as Reichenbach also admitted, Poincaré's claim is mistaken. Whatever rod deformations we postulate, if our measurements with some rods showed a non-Euclidean topology, they would do so with any rods subject to continuous deformation relative to them. Hence Reichenbach claimed that in order to postulate a geometry with a different topology we would have also to postulate 'causal anomalies'. This point I shall discuss shortly. But to return to metrical geometries of the same topology, as for instance are hyperbolic and Euclidean geometries, what shall we say of the claim of both Poincaré and Reichenbach, that, because we can postulate any deformations we wish, it is a matter of convention what geometry we ascribe to Space? Our answer must be that since, as shown in Chapter 4, it is not a matter of convention whether or not we postulate deformations, it will not be a matter of convention which geometry we ascribe to Space. However, we admitted in Chapter 4 that a situation might arise where there were two or more alternative corrections which could be made to measuring rods and no one correction be the right one. In that case there will be alternative ways of measuring Space. The different measuring procedures may or may not yield different geometries. If they do yield different geometries, we cannot give a simple answer to the question what is the geometry of Space. We can only say that if we measure in one way it is (e.g.) Euclidean, and if we measure in another way, it is (e.g.) hyperbolic; and that there is no reason for choosing one way of measuring rather than another, for both ways are equally satisfactory ways of explicating our ordinary concept of distance.

Physical Space, we urged (p. 26), of logical necessity can have no boundary. But this does not show that it must be infinite. For it may be the three-dimensional analogue to the surface of a sphere, which, though of finite area, has no edge. If in physical Space all the straight lines are closed lines, returning to their starting point, Space will be finite. For in that case there will be only so many regions of the Space. A finite or closed space (I use the term 'closed' as equivalent to 'finite without a boundary' and thus in the cosmologist's and not the topologist's sense) is one in which there are only a finite number of regions of finite volume excluding each other. The volume of such a space is the sum of all such regions which can be delineated. An infinite or open space is one in which there are an infinite number of regions of finite volume excluding each other. There are various open and closed 'spaces' discussed by geometers, some of which have different topologies from others, but all closed spaces, in our sense, have different topologies from open spaces. The space of elliptic geometry (like the space of spherical geometry, see p. 103 note 2) is a closed space.

That Euclidean space has a different topology from elliptic space can be seen as follows. Take any two points *P* and *Q* lying one on each side of an infinite Euclidean plane *E*. It is not possible for a line to join *P* and *Q* unless it passes through *E*. To whatever continuous deformations *E* is subject, the same will apply. However it is contracted, expanded or bent, so long as *P* lies on one side of it and *Q* on the other, any line joining *P* and *Q* must pass through *E*. Now in elliptic space no surface which is not an enclosing surface could have this property. By an enclosing surface I understand one topologically equivalent to the surface of any body such as a sphere which can be contracted to a point. Now *E* is not an enclosing surface. Hence *E* is topologically distinct from any surface in elliptic space, and hence Euclidean space has a different topology from elliptic space. On the other hand hyperbolic space has the same topology as Euclidean. *E* and a hyperbolic plane have the same topological properties, such as separating *P* and *Q* without being enclosing surfaces.

To say that a region of Space is finite is to say that measuring rods of some finite length can be laid between any two points of the region in a finite number of steps. To say that Space is finite is to say that there are only a finite number of regions of finite volume excluding one another, and so that there is a finite region which is closed. If the enclosing surface defining a finite region were completely surrounded externally by places belonging to the region, then the region would be a closed region — it would be all the space there was, and by exploring it we would have discovered what was the topology of Space. If two observers can reach all the points of a region enclosed within some boundary they will agree that it is finite. Once they agree that a region is finite, they will reach the same conclusions about its topology, whatever measuring rods they use. Whether they explore with an iron rod or an elastic band they would reach the same conclusion about, for instance, whether any non-enclosing surface of the region could separate any two points of it. For they would both reach all the places in it and discover which places were contiguous with which other places.

If we find by exploration a closed finite region, or have evidence that there is such a region, we may conclude that Space is finite, that that region is all the space that there is. If on the other hand the region which we have examined proves not to be closed, we cannot find out much about the topology of Space as a whole merely by extrapolation from its topology. We can only reach conclusions about the topology of Space as a whole by extrapolation from the metric of the finite region. This is because extrapolation from the known to the unknown does not work in quite the same way for topologies as for metrics and for the following reason. Having found that the metric of a small region of space is M_1, we make the simplest hypothesis about every other small region of space that its metric is M_1 and thence can deduce that the metric of Space as a whole is of a certain kind. (It is true that we can only measure the metric of a region of Space to a certain degree of accuracy, and that more accurate measurements might give a different result. Measurements on Earth with instruments known to the ancients showed that the metrical geometry of Space near the surface of the Earth was Euclidean, but more accurate measurements today might yield a different result. Nevertheless we must in any stage of science rely on the measurements available.) For topologies too, when we have found that the topology of a small region of Space is T_1, we may make the simplest hypothesis about every other small region of Space that its topology is T_1. But even if we ascertain (and ascertain in fact correctly) that the topology of every small region of Space is T_1, we still do not know the topology of Space as a whole, for there are often many different topologies of Space as a whole having topology T_1 in every small region.

This point can easily be brought out with a two-dimensional example. Suppose we find that a surface in three-dimensional Euclidean Space has in a small area a metric M_1 (such that [e.g.] the interior angles of all triangles exceed 180° by a certain amount). Then if we assume that every small area of the surface has the metric M_1, it follows that the surface is the surface of a sphere. But if we know that the surface has the topology T_1 (say, the topology of a disc) in a small region and assume that it has everywhere, we still do not know what is the topology of the surface as a whole. The surface may be a plane or the surface of a sphere — in every small region these have identical topologies. It should be added that although many different topologies of Space are compatible with the assumption that in every small region the topology is the same topology T_1, this assumption does rule out many different topologies of Space. But it does not enable us to choose, for example, between Euclidean and elliptic geometries which have in every small region the same topology. Whereas the assumption that every small region of Space has a metric M_1 does uniquely identify the metric of Space as a whole. Hence if we are to extrapolate from properties of an explored region to find the topology of Space as a whole, we must use its metrical properties.

In his discussion of the question of what we should say about the topology of Space if the region explored appeared closed, Reichenbach ([7] § 12) urged that we could always consistently deny any conclusion based on exploration with certain measuring rods about the topology of Space by postulating causal anomalies. He considers a case where a man leaves his study, passes through a number of spherical shells, each completely enclosing the next one, and finds himself in a room exactly like his study. There are then, Reichenbach urges, two possible interpretations of what has happened. *Either* geometry is non-Euclidean (e.g. the torus geometry postulated by Reichenbach), so that you can pass through spherical surfaces enclosing each other and find yourself back at your starting point, *or* the place where the man found himself was not really his starting point. The room was merely one very like his study. Reichenbach's explorer does an experiment to see if the new room is really his old study. 'He writes down his thoughts on a sheet of paper, adds a code word, locks the paper in a drawer, puts the key in his pocket, and leaves shell 5. . . . Arriving at 1 he finds his room, opens the drawer with the key he put into his pocket, and recognises on the slip of paper the same words which he had written down in shell 5' ([7] p. 64). Here, claims Reichenbach, the observer can either take this evidence as evidence that the two places are the same place or postulate action at a distance between the two places, so that his writing something down in one place produced a similar piece of writing in the other place without intervening places being affected (viz. there being no known wireless devices or such transmitting through the shells and thus producing the effects). Now, claims Reichenbach, such action at a distance is to be considered a 'causal anomaly' of the type that science attempts to avoid. We can, he says, still claim that geometry is Euclidean but to do so we would have to postulate causal anomalies, and this he advises us not to do.

Two points must however now be made against Reichenbach's claim about his parable. The first is that, contrary to what Reichenbach seems to suppose, there is a truth here. One of the above interpretations is true and the other false, whether or not the observer could come to know which is which. We can see this by reflecting on the fact that it is at any rate logically possible (and perhaps physically possible) that the investigator could grow in size until his arm reached right through the shells. Let him then touch one study with one hand and reach through the shells with the other arm. If the geometry was Euclidean, he would touch a different study with the second arm. If it was non-Euclidean, he would touch the same study and hold the first hand with the second. The two situations are different. Secondly even if an observer could not grow in size, he still has good reasons for a belief that the topology is non-Euclidean. The first good reason is that as we have seen, simplicity is evidence of truth, and the fact that it is necessary to postulate an extra and strange force producing duplication of effects at a distance in order to save Euclidean topology is good reason for holding that the topology is non-Euclidean. The second good reason is that, as we saw in Chapter 1, the detailed similarity of a present object to a past object is evidence that the two objects are the same object — in the absence of positive evidence that they are not. The fact that both studies are qualitatively similar and surrounded by qualitatively similar objects in each case, as far as you go from the two studies, is strong evidence that the two studies are the same material object, and so that the topology is non-Euclidean.

The general point which seems to lie behind Reichenbach's use of his parable is that one can 'choose' whether to say that two objects are or are not the same object. Once a decision has been made, with respect to all objects linked by paths, which objects are identical to which other objects, the topology of space is fixed. The metric is then determined by which paths are straight lines, and this, I urged earlier, is not in general an arbitrary matter. I now urge that the topology is also no arbitrary matter. There is a truth as to whether or not objects are identical, and one about which we can have good evidence. That evidence of qualitative similarity is evidence of identity is, as we saw in Chapter 1, a basic criterion for our interpretation of experience.

All of this provides a conclusive objection to the Kantian position that Euclidean geometry is the only geometry which, it is logically possible, can hold of physical space. For the topology of Space might be non-Euclidean, and observers might have good reason for believing that, which they could only deny by denying that evidence of qualitative similarity is evidence of identity. And if we denied that, as we saw in Chapter 1, we would never be able to make any justified judgements at all about the identity of material objects.

It follows from this that if a cosmological theory claims that Space measured by certain rods has an overall positive curvature, we cannot reinterpret it so that it claims that Space is Euclidean but that 'causal anomalies' occur. Hence what I have called in Chapter 4 (pp. 77ff) the two interpretations of the General Theory of Relativity are in fact two different theories making two different sets of predictions. The original interpretation was that the geometry of Space was Euclidean, but that all rods were deformed near massive bodies. The later and now traditional interpretation is that very short rods of negligible thickness preserve everywhere the same length, but that the geometry of Space is affected by the presence of massive bodies, being, Einstein argued in 1917, on the cosmic scale of positive curvature. On a smaller scale both theories make the same predictions, but on the cosmic scale — given Einstein's 1917 argument — they do not. For the later interpretation claims that the topology of Space is non-Euclidean, that Space is closed, while the former interpretation claims that it is Euclidean. The original interpretation claims that all rods are deformed by the presence of massive bodies, but, as we have seen, however they are deformed, they will measure the true topology of Space. Hence this interpretation predicts that measuring rods will measure a Euclidean topology, while the later interpretation claims that they will measure a non-Euclidean topology. The grounds for choosing between the two interpretations are the normal grounds for choosing between scientific theories. I shall henceforward understand by the General Theory of Relativity the interpretation which involves the claim that very short rods of negligible thickness preserve everywhere the same length, since this is the one normally used by physicists today.

In the light of the argument of this chapter, it is now clear how the question raised at the end of Chapter 3 whether Space is M-absolute or M-relative is to be settled. To say that it is M-absolute is to say that its geometrical properties are independent of the physical objects which it contains; to say that it is M-relative is to say that they are dependent on them. If the geometry of Space varies from place to place, dependent on the concentration of physical objects, then Space is M-relative; if it does not, then Space is M-absolute. The geometry of Space is the geometry which would be measured by a rod corrected for all deforming influences. To account for coincidences between approximately rigid rods not being preserved under transport, we have to postulate differential influences. If it adds to the simplicity of the physical system, we must also postulate universal influences, so long as our ordinary understanding of when a rod has approximately preserved its length is not thereby violated.

In this chapter we have considered what it means to say that Space has a certain geometry. The Space in which we live is clearly far too large for investigations into its geometry to be carried out in terms of rigid rods. The question therefore arises how we can determine by less direct methods the geometry of our Space. This question will be considered in Chapter 14.

BIBLIOGRAPHY

[1] Euclid, *The Elements*.

For an elementary summary of the properties of the three non-Euclidean congruence geometries discussed in the text, see:
[2] H. P. Manning, *Introductory non-Euclidean Geometry*, New York, 1963.

For a good historical account of the genesis of non-Euclidean geometry, see:
[3] R. Bonola, *Non-Euclidean Geometry* (trans. H. S. Carslow), New York, 1955.

For Einstein's original presentation of the General Theory of Relativity and his original conclusions about the geometry of space, see:
[4] A. Einstein, 'The Foundation of the General Theory of Relativity' (originally published 1916) and 'Cosmological Considerations on the General Theory of Relativity' (originally published 1917) in A. Einstein *et al.*, *The Principle of Relativity* (trans. W. Perrett and G. B. Jeffery) London, 1923.

For philosophical discussion of the issues see:
[5] I. Kant, *Critique of Pure Reason*, Transcendental Aesthetic.
[6] H. Poincaré, *Science and Hypothesis* (originally published 1902) New York, 1952, part ii.
[7] H. Reichenbach, *The Philosophy of Space and Time* (originally

published 1928) (trans. M. Reichenbach and J. Freund), New York, 1957, ch. 1.

[8] Rudolf Carnap, *Philosophical Foundations of Physics*, New York and London, 1966, part iii. (This gives a simple modern exposition of Reichenbach's 'conventionalist' standpoint.)

7 The Dimensions of Space

The pure geometer generalising Euclid's system constructs self-consistent geometries of any number of dimensions.[1] Which of these can be physical geometries, viz. can have application to physical space when 'point', 'line', 'plane', etc., have the senses described in Chapters 2 and 6?

An important example of a non-physical interpretation of four-dimensional pure geometry is provided by the geometrisation of the physics of Relativity Theory initiated by Minkowski. His four-dimensional space is usually called 'Space-Time'. The points thereof represent point-instants, points of space at temporal instants. To give the point-instant of an event is to say where and when it occurred. To identify a point-instant four coordinates are necessary, viz. three spatial and one temporal. 'Space-Time' has become a central entity in the structure of the General Theory of Relativity essential for giving explanations and making predictions. But the 'distance' between two points in Space-Time is not a distance, nor the 'direction' a direction in the normal sense.

But we are concerned with the question whether in the normal physical senses of the terms it is logically possible that there be spaces of dimensions other than three, which are, of course, the dimensions of our own physical space. One- and two-dimensional geometries certainly have application to segments of our physical sense. But is it logically possible that there be a complete space of one or two or more than three dimensions?

Before answering this question we must examine the prior question what it means to say that a 'space' in the pure geometer's sense has, for some n, n dimensions. The normal way of answering this question until the beginning of this century and still subsequently provided in works on metrical geometry, though not in works on topology, is to say that a space is n-dimensional if and only if n real coordinates are necessary and sufficient for unique identification of points. Thus a Euclidean plane will

---

[1] For description of how multi-dimensional Euclidean geometries are developed from three-dimensional Euclidean geometry see [2].

be two-dimensional because if you mark a straight line on the plane, you can uniquely identify any point by its perpendicular distance from the line and the distance along the line to where the perpendicular from the point cuts the line, but not if you use a smaller number of measurements. If a point is *m* units of perpendicular distance away from a specified line and the perpendicular cuts the line *n* units along it, then with this system of identifying points it will have the coordinates (*m*, *n*). Two and only two seem to be needed. Physical space is three-dimensional because any point can be uniquely identified by its perpendicular distance from each of three planes; and thus apparently by three and no less than three coordinates.

The trouble with this definition, as it stands, is that by it all 'spaces', including our own physical space, would in fact turn out to be one-dimensional. This is because of Cantor's proof that all the pairs of rational numbers can be put into one-one correspondence with the rational numbers; and all the pairs of real numbers can be put into one-one correspondence with the real numbers (and so any plane into one-one correspondence with any line). Hence any *n*-ad of real numbers sufficient for unique identification of a point can be given a unique single number. One coordinate will suffice to identify any point in physical or any other 'space'.

Hence topologists favour a different kind of definition of dimension. Their definition adumbrated by Poincaré, and developed by Brouwer, Urysohn, and Menger[1] defines a space as *n*-dimensional if *n* is the least integer for which every point has arbitrarily small neighbourhoods whose boundaries have dimensions less than *n*. The dimension of a space is thus defined by the dimension of the neighbourhood of a point of the space, and it by the dimension of its boundary. The empty set is defined as having the dimension — 1, and hence a point has a dimension 0. Physical space is thus three-dimensional because any small region *R* of that space is bounded by an object *S*, and that object is bounded by an object *L*, and *L* is bounded by a point. So *L* must be one-dimensional, *S* two-dimensional, and *R* (and so physical space) three-dimensional.

Now I believe that the older definition can be made serviceable by placing an obvious restriction on the type of coordinate used, so that the Cantorian objection no longer applies. The restriction is that the coordinates shall be measurements of 'distance'. (If *n* measurements of 'distance' serve for unique identification of points, then *n* measurements

---

[1] For precise modern formulation of Menger's definition see [1], p. 24. For the history of the definition see [1] Chapter 1.

of 'distance' or 'direction' will also serve, since a measurement of a 'direction' is a measurement of the ratios of 'distances'.) This restriction is evidently tacitly implied in claims about the dimensionality of physical space which relied on the original definition. To say that physical space is three-dimensional is to say that three measurements of distance are necessary and sufficient to identify any point. If we replace the triad of numbers which identify a point by a single number, that single number cannot measure the number of units of distance (or direction) at which the point lies from parts of some frame.

This can be seen by a feature of the Cantorian method for putting the points of a plane P and a line L into one-one correspondence. The points of P and L can be put into one-one correspondence, but not into one-one continuous correspondence. This means that in general the straight lines of P will not correspond to continuous segments of L. Hence all points of P which are very close to each other can have associated with them pairs of coordinates, each member of which differs but slightly from the corresponding member of the other pair. But if the points are given single coordinates by being associated with L, this will not in general hold. Very close points will have wildly different coordinates. The points with the single coordinates 2 and 1 might very well be at far greater distance from each other than the points 1 and 1000. If single coordinates are used for identification of points of a Euclidean plane, they cannot be measurements of distance.

The older definition with the restriction runs as follows: a 'space' is n-dimensional if and only if n measurements of 'distance' (or 'direction') are necessary and sufficient for unique identification of any point. With the restriction the older definition and the topologist's definition will yield the same dimensions as each other for Euclidean spaces and many other spaces including the elliptic and hyperbolic spaces discussed in the last chapter, viz. if such a space is n-dimensional on one definition, it will be n-dimensional on the other.

Which definition are we to choose? Clearly, that definition which brings out best what we ordinarily mean when we say that a Euclidean plane has two dimensions or that physical space has three dimensions. Since the topologist's definition was only developed in the twentieth century[1] and the other definition is much older and is the only one which most men learn when they learn geometry, it would seem that the older

---

[1] The same is true of any other definitions which may be proposed, e.g. that provided in [1] p. 24, and hence the argument which I give against the topologists' definition, if valid, holds against them too.

definition is a better elucidation of our ordinary usage. This suggests that if the older definition and the topologist's definition were for some space to yield different results, we would say that the dimensions of the space were those yielded by the older definition. True, the original definition as amended is only of use for a metric 'space', viz. one in which an analogous concept to that of distance can be introduced, and the topologist's wider definition is of more use for his own purposes. But in dealing with the dimensions of physical space, we are of course dealing with a metric space.

I would suggest then that to say that a space has n-dimensions is to say that it is possible uniquely to identify every point by n measurements of distance or direction from some frame of reference and not possible by less than n measurements. For many metrics including a Euclidean one, the following system of measurement will give unique identification of any point P. In a one-dimensional space you need only specify distance from a given point along the line. In a two-dimensional space you need one straight line given in advance. A straight line is then drawn from P to the given line perpendicular to it. Distances are then measured from an origin to the point of intersection of the lines, and from the point of interesection to P. In a space of three dimensions a line is drawn from P perpendicular to a given plane. Two measurements of distance are then needed to locate the point on the plane where it is cut by the line from P; and another measurement to measure distance along the line from the plane to P. So three measurements of distance uniquely identify P. In a space of four dimensions, a line is drawn from P perpendicular to a given hyperplane, a hyperplane being a three-dimensional entity produced by moving a plane kept parallel to itself in a direction perpendicular to itself. Three measurements are necessary to locate the point of perpendicular intersection with the hyperplane of a line from the point, and a further measurement to measure the distance along the intersecting line. Four measurements of distance in all are thus necessary uniquely to identify a point; and so on for spaces of higher dimensions. We can always substitute a measurement of direction for a measurement of distance. Thus spherical polar coordinates use one measurement of distance (the distance along the straight line joining the origin O to the point under investigation P) and two measurements of direction (the angle which this line makes with a given plane, and the angle which a certain projection of the line on to the plane makes with a given line).

In a different metric from the Euclidean it might be necessary to make measurements from a frame of reference in a different way. Thus in a plane of some more complicated metric we might need to make one measurement along a given circle and the other measurement along a line of a certain curvature joining the circle and the point whose location we are measuring. We might need to do this because the space might be such that no line from certain points interesected at right-angles a given straight line such as we would use for measurements in a Euclidean space. But to claim that the space is n-dimensional is to claim that a way of measuring can be laid down in advance so that every point can be uniquely identified by giving n coordinates which represent measurements of distance and cannot be uniquely identified by giving less than n coordinates.

Clearly the dimension of a space, as we have defined it, is a property thereof invariant under continuous deformation and hence a topological property. For suppose an n-dimensional space S to be changed into a space S' by a series of continuous deformations D, D', D'', etc. Now if by measuring along certain lines n measurements were previously necessary, then n measurements will now be necessary if we measure along the lines obtained from the previous ones by the deformations D, D', D'', etc. Hence if n was the minimum number of measurements necessary for unique identification of points in S, points in S' can be uniquely identified by n measurements. Hence continuous deformation cannot increase the dimensions of a space. Nor therefore can it decrease the dimensions of a space; for, if it could, the reverse deformation would increase the dimensions.

Having now clarified what it means to say of a 'space' in the pure geometer's sense that it has, for some n, n-dimensions, we must now revert to applied geometry and inquire whether it is a logically necessary truth that physical space is three-dimensional.

Kant claimed 'that complete space (which is not itself the boundary of another space) has three dimensions, and that space in general cannot have more, is built on the proposition that not more than three lines can intersect at right-angles in a point. This proposition cannot be shown from concepts, but rests immediately on intuition, and indeed because it is apodictically certain, on pure intuition *a priori*' ([4] § 12). All Kant's philosophical predecessors who discussed the question also argued for the necessity in a stronger sense than 'physical' of the tridimensionality of space. Aristotle simply stated that bodies were divisible in three ways and that nothing was divisible in more than three ways.[1] Ptolemy, according to Simplicius, argued that space must be three-dimensional, because distances are measured along perpendicular lines and that only

---

[1] Aristotle, *On the Heavens*, 268a–b.

three lines can be mutually perpendicular at a point.[1] Ptolemy apparently was the first to use this argument, used afterwards by others including Galileo[2] and Kant[3]. They argued that the conclusion that space is n-dimensional follows from the premiss:

1. n and only n straight lines can be mutually perpendicular at any point.

However the conclusion only follows if further premisses hold. The two which these writers take for granted are:

2. Each point P can be uniquely identified by its perpendicular distance from some point P' on any object formed by moving one of the n lines perpendicular to each other at some point parallel to itself in a perpendicular direction, by moving the resulting surface parallel to itself in a perpendicular direction, and so on until $(n-1)$ lines are exhausted. P', given that it lies on the object previously described, can be uniquely identified by its perpendicular distance from a point P'' on any object formed by the same process from all but one of $(n-1)$ lines used previously, and so on. None of these points, however, can be identified by their perpendicular distance from objects formed by similar motions of a smaller number of perpendicular lines than those stated. Hence n measurements beginning with distance along a line from an origin point will uniquely identify P, but $(n-1)$ will not.

3. There is no other method of measurement whereby fewer measurements are needed uniquely to identify a point.

---

[1] Simplicius, *Commentary on Aristotle's 'On the Heavens'*, 7a, 33. Quoted in [2], p. 1.

[2] Galileo Galilei, *Dialogue Concerning the two Chief World Systems* (trans. S. Drake), (Berkeley and Los Angeles, 1962), p. 14.

[3] In his discussion Kant [4] also refers to the fact that similar but incongruent two-dimensional counterparts, viz. a shape and its mirror image, can be made congruent by rotation in a three-dimensional space; but similar but incongruent three-dimensional counterparts, such as a left and a right hand, cannot be brought into congruence by any known physical procedure. It was subsequently proved in the nineteenth century that any two $(n-1)$-dimensional objects metrically symmetrical about an $(n-2)$-dimensional object can only be brought into congruence in an n-dimensional space. The fact that the left and right hand cannot be brought into congruence by any known physical procedure is only further evidence for and consequence of the tridimensionality of physical space, not a proof of the logical necessity or contingency of this fact.

120   *Space and Time*

Both these latter conditions hold in Euclidean geometry if the first condition holds, but this is not so in all pure geometries. Given the latter conditions and given that only three lines can be mutually per-pendicular, it follows deductively that space must be three-dimensional. But to prove that the conclusion holds of logical necessity, we need first to show that geometries in which the two latter conditions stated above do not hold when the first does hold are not logically possible geometries. And secondly we need to prove that granted that in fact in our Universe no more than three straight lines can be mutually perpendicular at a point, this must be so of any universe. The logical necessity of these conditions seems no more obvious than that space be three-dimensional. Ptolemy has shown no firmer foundation for our proposition.

In a different tradition some seem to have reasoned somewhat vaguely that since for any n, there is a perfectly consistent n-dimensional pure geometry, the fact that points of physical space need three and only three coordinates for unique identification is a merely empirical matter. This of course does not follow, for it would have to be shown that there is nothing in the nature of a point, in the sense elucidated in Chapters 2 and 6, which limits its necessary identifying coordinates to three. No proposition about the possibilities of pure geometry could prove that. What is needed is an examination of the properties of points.

With the arrival of multi-dimensional geometries in the nineteenth century and even earlier, a number of attempts were made to show that the tridimensionality of space was contingent by attempting to explain it as the consequence of the operation of some scientific law. For if it can be shown that the tridimensionality of space is a consequence of some scientific and so logically contingent law, then if a different law held space would have different dimensions. Since any given scientific law might, it is logically possible, not have held, the dimensions of space would be a logically contingent matter. The prototype of such expla-nation was, interestingly, given by Kant in a phase of his thought prior to the writing of the *Critique* and the *Prolegomena*. He noted that 'substances in the existing world so act upon one another that the strength of the action holds inversely as the square of the distances' ([3] § 10). Gravitational effect is propagated in inverse proportion to the square of distances of bodies affected; and, as was discovered a few years after Kant had written this, so too is magnetostatic and electrostatic effect. Now let a source of attraction or repulsion be surrounded by particles all lying at some specified distance from it and completely enclosing the source. Let us call the sum of the changes of momentum produced by the source in all these particles at the distance the total effect of the source at that distance. If we assume that the total effect of some source of force is the same at any distance, and if we assume a Euclidean geometry, as Kant of course did, then in a three-dimensional space an inverse square law of attraction or repulsion must apply. For the particles completely enclosing the source all lying at some specified distance from it will form a two-dimensional surface, the area of which is proportional to the square of the distance from the source. Conversely if the attraction or repulsion on any particle is inversely proportional to the square of the distance, and the total effect at any distance is the same and geometry is Euclidean, space must be three-dimensional. Though Kant does not making his argument completely clear, it seems to be as I have stated it, and others[1] used later the argument which I have stated. This argument is perfectly valid, but the conclusion which the early Kant wished to draw from its conclusion — that the tridimensionality of space was logically contingent — does not follow. Kant claimed that owing to the inverse square law 'the whole which thence arises has the property of threefold dimension, that this law is arbitrary, and that God could have chosen another, for instance the inverse threefold relation; and lastly that from a different law an extension with other properties and dimensions would have arisen' ([3] § 10). This further argument is that since the existence of inverse square laws as opposed to inverse cube laws is clearly contingent, then if we suppose a different law to hold, e.g. an inverse cube law, given also that the total effect of a force at any distance from a source was the same and space was Euclidean, space would have other dimensions. Hence the tridimensionality of space is a contingent consequence of the existence of inverse square laws. But although the existence of inverse cube laws is clearly logically possible, the question arises whether the existence of inverse cube laws is logically compatible with the operation of the principle of total effect in a Euclidean space. Only if it could be shown that the three suppositions together describe a logically possible universe, could the tridimensionality of our space be attributed to the operation of inverse square laws. Only then would it follow that God could have brought about 'an extension with other properties and dimensions'. The later Kant no doubt saw this point.

Similar arguments purporting to prove that the tridimensionality of space was the consequence of some scientific principles and so contingent have been much used ever since Kant.[2] They mostly have the

[1] See [7] p. 177, for examples.
[2] See [7] for an account of these.

same structure and suffer from the same deficiency as his argument. They have two premisses — one P_1 describing a purported contingent feature of our Universe (in Kant's example, the operation of inverse square laws) and the other P_2 (sometimes not made very explicit) a proposed necessary truth. It may be alleged that P_2 states a necessary truth in the ordinary senses of the terms (as Kant seems to have considered that the combined principles that the total effect of a force is the same at any distance from a source and that the geometry of space is Euclidean, did). Alternatively it may be proposed that P_2 be taken as stating a necessary truth, be adopted as a useful convention for science. But then given that P_2 is necessary and P_1 contingent, 'P_2 and not — P_1' must describe a possible state of affairs. For no necessary truth can rule out some contingent possibility. Hence, the argument goes, the dimensions of space depend on the contingent feature of our universe described by P_1 and so are a contingent matter. On investigation however, it usually proves that P_2 does not state a necessary truth, in the ordinary senses of the terms. If we admit that P_2 states only a contingent truth, it is open to question whether 'P_2 and not-P_1' describes a logically possible state of affairs; for if we accept P_2 we may be committed thereby to denying not-P_1. If on the other hand P_2 is a necessary truth or we take it as such, it becomes in no way obvious that P_1 is a contingent truth, that the opposite of P_1 is at all conceivable.

I will take just one example of another such argument to illustrate the fallacies. This argument to prove the contingency of the dimensionality of space comes from Reichenbach [3], and is in fact a more general form of Kant's argument. His P_1, which he does not give in detail, would consist of a full description of how events in our universe at one place affect events at first at a certain group of places, and then another group of places; of how effects are propagated. P_2 is the proposition that all action is by contact, which, it is claimed, would be a useful proposition for science to adopt. Now P_1 is superficially contingent. If I press a button on my desk, what happens where next is a logically contingent matter. Perhaps a light will go on in the ceiling or a current be produced in an electric wire or anything may happen. If, however, we assume that P_2 states a logically necessary truth it follows that the places affected by my action must be contiguous with the button, and this with the information provided by P_1 leads, Reichenbach claims, to a unique dimensionality of space. Thus if the places are not contiguous in a three-dimensional space, it follows that there is a further dimension. Hence, the argument goes, the dimensions of space are a consequence of the contingent truth stated by P_1.

Now in fact P_2 in the ordinary senses of the terms does not state a necessary truth. It clearly makes perfectly good sense in an ordinary way to suppose that my action affects a distant body immediately without affecting the intervening medium. We can imagine this happening; we can imagine that every test to prove that the intervening medium was affected in intermediate time failed. On the basis of this we would ordinarily be regarded as justified in concluding that actions could have effects at a distance without the intervening medium being affected. If, however, we insist that in some way not directly detectable the medium must have been affected, it immediately becomes doubtful to what extent P_1 is contingent. It is no longer obvious that it is contingent what places I can immediately affect by my action. Kant's, Reichenbach's, and similar purported proofs of the contingency of the tridimensionality of space fail in the same way.

Admittedly, as I argued on p. 7 the fact that evidence appears to confirm a statement gives grounds for supposing that that statement is a meaningful factual statement. Evidence appears to confirm a statement if that statement is simple and successfully predicts that evidence. It might be possible to argue (as I do not think that Kant and Reichenbach effectively do argue) that some physical theory including a claim that Space was n-dimensional (where n is not equal to 3) was a simple theory which would lead us to expect certain observational evidence, which we do in fact find; and hence that there could be evidence which apparently confirmed the claim that Space had dimensions other than three. But such evidence would only really confirm that claim if the claim was a logically possible one, that is if it contained no contradiction. Nonsense may sound simple and predict successfully. I shall suggest that there are positive grounds for believing that the claim that physical Space has dimensions other than three is not a logically possible one. I shall do this by attempting to describe worlds of dimensions other than three. We shall find difficulties of different kinds standing in the way of giving coherent descriptions of worlds of less than three dimensions, and of giving coherent descriptions of worlds of more than three dimensions.

The places of such worlds will be of dimensions other than three and so too therefore will be the 'material objects' which occupy them. However on the analysis which we gave of a material object in Chapter 1 'material object' was defined by some standard examples, rocks, chairs, tables, etc., and whatever is composed of such, composes such, or excludes such. All of these examples being apparently three-dimensional, it looks as if on this analysis material objects must be three-dimensional, and so must be the places which they could occupy. In my investigations I shall for the present ignore this point and assume that a wider understanding of 'material object' is possible.

Recalling then that to say that Space is n-dimensional is to say that n measurements of distance (or direction) are necessary and sufficient uniquely to identify any point, let us see if we can give a coherent description of a two-dimensional world.[1]

In a two-dimensional world 'material objects' including agents would be two-dimensional, like patterns or paintings or the surfaces of objects in our world. They could be moved and, if agents, move themselves in all directions across the two-dimensional surface, which was the Universe, unless impeded by other objects or by forces. The boundaries of material objects would be formed by lines, and an agent touching objects or perceiving them in other ways would perceive their one-dimensional boundary and thereby learn about the objects. He would not be able to see the inside of objects without looking through their one-dimensional boundaries (similar to edges in our world). We may also suppose — to satisfy Reichenbach — that effects would be propagated with finite velocity by contact and only across the surface. In such a world two measurements of distance would appear to suffice uniquely to identify any point; for one and only one point P would lie n units of length along a line of a certain curvature intersecting a fixed line L at a certain angle, the intersection with L of the line from P lying m units along L from an origin.

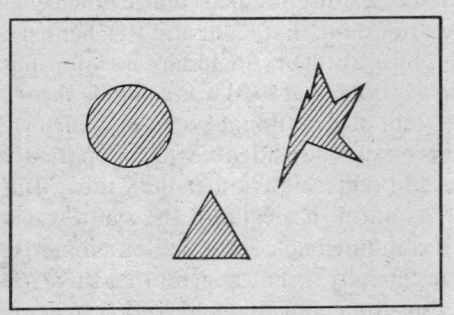

Figure 6
Diagram of 'material objects' in a 'two-dimensional world'

Agents in the world which I have described might well conclude that space is two-dimensional. We have supposed that it is physically impossible for material objects to move or effects to be propagated except on the surface. Hence inhabitants of such a world *might* well have no conception that it is logically possible that the objects should be moved otherwise than on the surface. They might have no such

[1] For a much fuller attempt to give a coherent description of a two-dimensional world, see the Victorian romance *Flatland*—[8].

conception. But it seems in no way necessary that they lack such a conception. For if they reflected on the structure of the 'boundaries' of 'material objects', as I have described them, it might occur to them that while the 'boundaries' had one very noticeable dimension of length, they also all possessed a very tiny height, and so it be possible for the 'material objects' to move out of the surface. This might in fact escape their notice, but I see no necessary reason why it should. Anyway, whether or not the inhabitants noticed the fact, it is clearly logically possible that the two-dimensional 'material objects' should be elevated above the surface or depressed below it. We, aware that patterns and pictures can be moved upwards as well as sideways, realise that the logical possibility exists even if the physical possibility does not. Since it is logically possible that the 'material objects' be moved out of the surface, there must be places, and so points, outside the surface, since a place is wherever, it is logically possible, a material object could be. Hence points would need a third coordinate, specifying their distance above or below the plane, to identify them uniquely. The only way in which we can describe 'material objects' of a two-dimensional universe is as patterns or pictures in our universe, but this very description permits the logical possibility of their moving above or below the surface. Hence the purported description of a two-dimensional world fails. The world described by me would in fact be a three-dimensional world.

The situation then in the two-dimensional case is that there is a perfectly conceivable situation in which the inhabitants of a universe could claim with *some* justification that space was two-dimensional, but we with superior knowledge of the possibilities of motion of material objects would know that they were mistaken. There cannot in fact be a two-dimensional space. What applies to the two-dimensional case must *a fortiori* apply to any alleged one-dimensional case. I therefore conclude that there cannot be a space of less than three dimensions.

What of space of more than three dimensions? Could there be a four-dimensional space? In such a world we would need four coordinates to determine points. If we set about locating points by the criteria of our world (e.g. perpendicular distance from a point on a given plane) we would find that very many points satisfied any description. We would find that very many points were x ft. of perpendicular distance from any point on some plane and no other system of identifying points by three measurements would serve for unique identification. However, four measurements would suffice to identify any point uniquely.

Material objects would also be four-dimensional and would be terminated by three-dimensional boundaries similar to our material objects. Objects and agents could move and be moved not merely or necessarily in the directions known to us (which are combinations of the three directions given by three rods meeting at right-angles to each other) but in infinitely many other directions. (In a four-dimensional universe of Euclidean geometry or many other geometries four rods could meet mutually at right-angles — see Figure 7. The possible directions of motion would then be combinations of these four directions.) Since material objects would be terminated by three-dimensional boundaries an observer would be able to see all points on these 'straight-off', and perceive four-dimensional objects through perceiving their three-dimensional boundaries.

Rods mutually at right angles in our world.

Rods mutually at right angles in the purported four-dimensional Euclidean world.

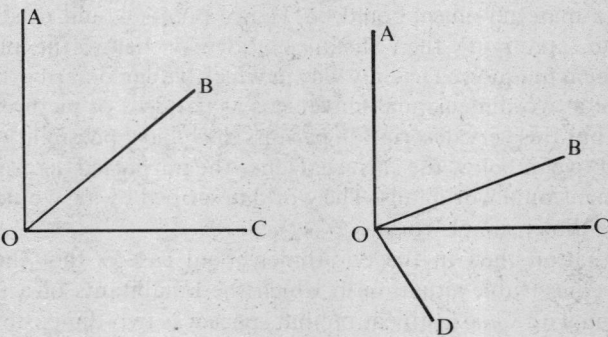

Figure 7

Such a world is, I now argue, inconceivable. If such a world were conceivable, then an inhabitant of it, if it were to exist, would judge that our world was merely the surface of a four-dimensional world, that the material objects of our world seemed to us to have only three dimensions but in fact had a fourth. Because it is not physically possible that our objects move in a direction other than the three familiar to us we — falsely, according to such an outside observer — would claim that no other type of movement was logically possible. But physical impossibility would have, he would say, closed our eyes to logical possibility. If there were a four-dimensional universe, this claim of its inhabitant would be correct. For his argument would have exactly the same structure as our argument to prove the impossibility of a two-dimensional world. He would be aware of logical possibilities to which our eyes are closed.

Now if our 'space' is in fact the boundary of a four-dimensional space, there must be directions in our Universe other than those which are combinations of the directions given by three mutually perpendicular rods. Where are these directions? An object should be able to move in directions other than back and forward, to this side and that, up and down. But how could it? The suggestion that it might does not describe anything which we can conceive, of which we can make sense. No motion is conceivable which takes us 'outside' our three-dimensional world and enables us to look at it 'thence'.

A second objection, similar to the first, is this. An observer outside our hyperplane would be able to observe at any instant all the 'inside' of any three-dimensional object known to us, such as a cube, just as we can observe at an instant the inside of any two-dimensional object. This initially seems conceivable. We imagine ourselves seeing the inside of a cube if we look into it through a number of transparent layers. But this is not seeing the 'inside' of the cube in the way that we see the inside of a surface. We can see the inside of a surface without looking through transparent layers of the surface from the edge. But what would it mean to see the 'inside' of a cube without looking through its bounding surfaces? No sense seems to pertain to such a notion.

Thirdly, if our world was a four-dimensional world, a man ought to be able to pass across a cube without passing across its two-dimensional bounding planes; as we can pass across a two-dimensional surface without passing over its one-dimensional edges. Yet no meaning would seem to attach to such a purported description. For these reasons I conclude that no sense can be given to the thesis that we live in a four-dimensional space and hence to the thesis that space could be four-dimensional. My objections apply *a fortiori* to the suggestion that space might have dimensions greater than four. I therefore conclude that space must of logical necessity have three dimensions.

Some have attenpted to make sense of a four-dimensional space by picturing as in Figure 7 four rods mutually at right-angles. We imagine ourselves measuring the six angles $\angle AOB$, $\angle BOC$, $\angle COD$, $\angle AOC$, $\angle AOD$, $\angle BOD$ and finding that they each measured a quarter of a circle. But even if we discover that n and no more than n rods fit mutually at right-angles, that does not by itself mean that we have found an n-dimensional universe. It will only do so, as we have noted on pp. 119, if further conditions are satisfied. We noted two conditions true in our universe, the joint satisfaction of which guarantees that if four and no more than four rods can meet mutually at right-angles, space is four-dimensional. It is not, however, logically possible that one of these conditions be satisfied if four rods meet mutually at right-angles. This is the condition which states that each point of space P can be uniquely identified by its perpendicular distance from some point P' on some object constructed by the procedure described on p. 119 from $(n-1)$ perpendicular lines but not from the object constructed by that procedure from $(n-2)$ perpendicular lines. This means that we would have to be able to construct a three-dimensional object at some point O by moving the plane OAB parallel to itself along the line OC, and measure distance along a line perpendicular to this object, viz. OD. Hence OD would have to pass through O without touching the object at any neighbouring point — for if it did touch the object at a point next to but distinct from O, it would not be perpendicular to the object. That OD touch the object at no neighbouring point is clearly inconceivable. Hence it is inconceivable that any point can be identified by its perpendicular distance from any object constructed from three mutually perpendicular lines by the procedure described on p. 119. So the further condition would not be satisfied. Even if four rods can be placed mutually at right-angles,[1] that does not guarantee a four-dimensional geometry.

The position then with purported two- (and one-)[2] dimensional spaces is that a certain world which many might wish to describe as a

---

[1] In the most convincing description of a four-dimensional universe known to me Honor Brotman [6] has argued that it is conceivable that four rods meet mutually at right-angles. However she has not shown this because in order to establish that the angles are right-angles (in her terminology 'square angles') she uses as a criterion for the straightness of a rod ([6] p. 254) 'The look of the thing'. But clearly a rod may look straight without being straight. The tests for the straightness of a rod are that it lies along a straight line, the tests for which I have described on pp. 63f. However, she may well be right that it is possible for four rods to meet mutually at right-angles. But I claim to have shown that if four rods could be arranged mutually at right-angles this would only mean that the Universe was four-dimensional if further sets of conditions were satisfied, and that the most likely of these of logical necessity cannot be satisfied if four rods meet mutually at right-angles.

[2] Though there are difficulties with a one-dimensional space which do not arise with a two-dimensional space. Given that — by definition — material objects cannot be penetrated by other material objects, no measurements could take place in a 'one-dimensional space' since 'material objects' could not there be circumvented by measuring rods. Measuring rods which are not material objects would not measure distance as we have analysed distance. So how could sense be given to the notion of something being at a certain distance from something else? Abbott's description of 'Lineland' in [8] sections 13–14 does not really meet this difficulty. He replaces the concept of distance by something bearing little resemblance to it.

two-dimensional world was conceivable but that the description was mistaken. For purported spaces of more than three dimensions the position is that the situation which we would wish to describe as a four-dimensional world is not a conceivable one. The concepts such as straightness and angle by which we determine the distance and direction of objects are such that it is not possible that a point could not be identified uniquely by three measurements of distance and direction. Of course one could so define concepts of 'straight' and 'angle' and 'place' that this was possible and so also that space be four-dimensional. But to say this is no more informative about physical space that it would be informative to be told about mathematics that with suitable definitions of '2', ' + ' and ' = ' one could ensure that '2 + 2 = 5'. With such concepts of straightness and angle as we have, space must be three-dimensional.

But, it may be objected, how do we know that we are not in the same situation as the inhabitants of the purported two-dimensional world described earlier? It seemed to them that only two lines could be mutually perpendicular. They made this mistake because it was not physically possible for the 'objects' with which they were familiar to move outside their surface. Might not we be making a similar mistake in supposing that our space is three-dimensional because it is not physically possible for the objects with which we are familiar to move outside the three-dimensional hyperplane? It must be admitted that we *might* be making just this mistake. But the concession that what we have apparently proved to be clearly inconceivable might, just might, be conceivable, is a concession that must be made, as similar concessions must be made for all purported proofs in philosophy and mathematics and in other fields. There might, just might, be a surface which was both red and green all over at the same time, despite the apparent inconceivability of such a surface. All the human race might be subject to some odd mental blockage in failing to realise that such a surface was a logical possibility. So too our proofs that a four-dimensional (or two-dimensional) space is inconceivable might be wrong. But I urge that the difficulties of giving sense to these notions are apparently overwhelming, and hence the onus is on him who wishes to deny our thesis to overcome these.

BIBLIOGRAPHY

For a modern mathematical exposition of the theory of dimensions, see:
[1] W. Hurewicz and H. Wallman, *Dimension Theory*, Princeton, 1941.

For the basic theorems of four-dimensional Euclidean geometry, and for a clear exposition of the way in which multi-dimensional geometry has developed from three-dimensional geometry, see:

[2] H. P. Manning, *Geometry of Four Dimensions*, New York, 1956.

For philosophical discussion of the necessity of the tridimensionality of Space, see:

[3] I. Kant, *Thoughts on the True Estimation of Living Forces* (first published 1747), §§ 9–10. A translation by N. Kemp-Smith appears in *Kant's Inaugural Dissertation and Early Writings on Space* (trans. John Handyside), Chicago and London, 1929.

[4] I. Kant, *Prolegomena to any Future Metaphysics* (first published 1783) (trans. P. G. Lucas), Manchester, 1953, §§ 11–13.

[5] H. Reichenbach, *The Philosophy of Space and Time* (originally published 1928) (trans. M. Reichenbach and J. Freund), New York, 1958, § 44.

[6] H. Brotman, 'Could Space be Four-dimensional?', *Mind*, 1952, reprinted in *Essays in Conceptual Analysis* (ed. A. Flew), London, 1956. My page reference refers to the latter volume.

For a history of attempted scientific explanations of the tridimensionality of Space, see:

[7] M. Jammer, *Concepts of Space*, Cambridge, Mass., 1954, pp. 172–84.

For a novel purporting to give coherent descriptions of one-, two-, and four-dimensional universes, see:

[8] E. A. Abbott, *Flatland*, 2nd and revised edn, London, 1884. Republished Oxford, 1962.

8 Past and Future

So far I have dealt almost entirely with Space and its properties while using temporal terms unanalysed. In this and the next three chapters I shall be concerned to elucidate the meaning of temporal terms and propositions about time.

Spatial things exist, their states persist or change during periods of time; and anything during which a spatial thing could, it is logically possible, exist is a period of time. Events are changes of states of things. Not all things, the existence, states, and changes of state of which can be dated, need have spatial location. As Kant pointed out, thoughts and feelings exist for periods of time, but do not exist at a place.[1] The class of temporal things is thus wider than but includes the class of spatial things. Yet, as we noted on p. 15, in order to refer to temporal things as to anything else we have to do so via spatial things. In order to identify a thought we have to identify the person, a spatial thing, whose thought it is.

A temporal instant or moment is the smallest bit of time. Temporal instants form the boundaries of periods of time. A period or interval may last, say, from 3.00 p.m. on Tuesday until 4.30 a.m. on Wednesday. To say that time is discrete or atomic or granular is to say that there are only a finite number of instants in any period bounded by instants. To say that it is dense or continuous is to say that there are in any period an infinite number of instants (either of the order of the rational numbers, or of the order of the real numbers. See p. 26 for an account of these numbers in the exposition of the similar claim about space.) The only meaning which I can give to such claims is as claims about the temporal intervals which events can last — e.g. to say that time is discrete is to say that there is some small finite interval such that it is physically possible that events last an integral number of such intervals, but not $1\frac{1}{2}$ or $2\frac{1}{2}$ etc. such intervals. In a similar way to the way in which Euclid defined a 'point' (see p. 99), Aristotle (*Physics*, 233G–234G) defined an 'instant', as a unit of time without temporal parts. If time is discrete in any sense,

[1] I. Kant, *Critique of Pure Reason*, B50.

its instants will last some time; but if it is not discrete, they will last no time at all. Still, even if time is not discrete, anything which lasts a very short period of time in comparison with the periods in which we are interested — e.g. the hand of a clock pointing to 12, or the sounding of a short note — may be regarded for practical purposes as occurring at an instant. The English word 'time' is unfortunately used in two very different senses. In the first sense a 'time' means a temporal instant. When we speak of the time at which events occur, we mean the temporal instant at which they occur. In the second sense a 'time' is the sum of temporal periods temporally related to each other. Periods are parts of time in this sense. Space is the whole made of places, very small places being points. But time unfortunately, in English, is made up of temporal periods, very small periods being times. Having drawn attention to the ambiguity, I shall avoid confusion by using 'time' in the second sense only. I shall use and have used 'temporal instant' to do the job done by 'time' in the first sense.

In order to identify temporal objects, their states and changes of state, uniquely, we have to specify, not merely where they existed or occurred, but when. One instant of time is the present one and other instants relative to it are past and future (I shall consider in Chapter 10 the thesis that of logical necessity there is only one time and consequently that all other instants are, relative to a given instant, past or future). Every past instant was a present one. Every future instant will be a present one. The present instant was future and will be past. All this is involved in what we mean by past, present, and future.

The most basic and important thesis about time from which I shall draw out most of the other logically necessary properties of time is the thesis that the past is determinate of logical necessity, that is, that — of logical necessity — the actions of agents and other causes can only affect present or future states (and so changes of state, events) and not past ones. I shall devote most of this chapter to discussing this thesis. In discussion of this and related questions in subsequent chapters I shall often refer only to states or only to events, viz. changes of state. But what I say about the one will apply to the other.

In one sense of course we can easily affect what is past and gone. What men do now may very well make a difference to the correct description of past events. Whether or not the atom bomb dropped on Hiroshima was the first of the only two ever to annihilate large cities depends on the subsequent actions of soldiers and statesmen. Merely by pressing a button a future statesman can make it not to have been the case that the Hiroshima bomb was the first of the only two ever to annihilate large cities. Again, whether or not Father Jones baptised in 1964 the next century's greatest pianist depends on the subsequent career of little John. But when philosophers and others claim that the past is unalterable, they are not thinking of these cases. They are thinking of cases such as the following. Nothing anyone can do now can make it not have rained yesterday, or me not have travelled by train on Monday. How can we distinguish the two sorts of case?

Statements apparently about a certain instant which, had they been correctly positioned and executed all logically possible tests, observers could have verified or falsified more conclusively at the instant referred to than at any subsequent or prior instant, are indeed statements really about that instant. Statements apparently about a certain instant which observers, had they been correctly positioned and executed all logically possible tests, would not have been able to verify or falsify at the instant referred to more conclusively than at a prior or subsequent instant, are statements with a covert past or future reference. By these criteria 'it rained yesterday' turns out to be a statement really about the past. No subsequent tests could show that it rained yesterday better than the best tests which could have been made yesterday. On the other hand 'Father Jones baptised in 1964 the next century's greatest pianist' turns out to have covert future reference. Subsequent tests could show its truth or falsity better than the best tests which could be made at the time (one could show its falsity at the time only if Father Jones baptised nobody in 1964, but this is a circumstance which need not hold. All that is necessary for the statement to have a covert future reference is that future tests *could* be more relevant to establishing its truth-value than any possible present tests.) Likewise '1972 will be the centenary of the passing of the Ballot Act' seems to be about a future year, but actually has a covert past reference, since verification and falsification could be achieved more conclusively in 1872 than in any other year.

The distinction of importance for present purposes is between statements really about the past and statements apparently about the past but possessing a covert future reference. Statements apparently about the past which turn out to have a covert future reference may in fact be partly about the past and partly about the future. Thus the statement made in 1968 'Father Jones baptised in 1964 the next century's greatest pianist' can be analysed into two components: 'Father Jones baptised at least one person in 1964', and 'One person baptised by Father Jones in 1964 will become the greatest pianist of the next century'. The first component is by our standards solely about the past and the second makes no pretence so to be (though of course it presupposes the truth of a past statement for the applicability of its referring term). Some statements apparently about the past but proving to have a covert future reference may not be about the past at all. Of this type is 'It was the case in 1964 that there will be a world war in 1984'.

The above distinction between statements really about the past and statements with a covert future reference is one on which I shall rely much in subsequent argument, and so it is important for me to make quite clear what I am claiming. I am not claiming that we cannot in practice often know subsequently (or prior) to some instant t_1 what happened at t_1 better than we knew at that instant. But I am claiming that for any amount of evidence we could have subsequently (or prior) we could (logically) always have better evidence at the instant. Whatever the evidence at t_2 about who killed Smith at t_1 still better evidence could be obtained at t_1 by galleries of observers of the killing, provided with evidence of the identity of the participants. It might not of course be physically or practically possible for an observer to make the necessary tests at t_1, to ascertain whether A occurred at t_1. Whether the temperature of the interior of the Sun was 5m°C at 4.00 p.m. on 1 January 1966 is not something which could have been ascertained by sending an observer to the middle of the Sun with some apparatus at that instant. This is because the only observers which we could have sent and who could have reported what the dial of the appropriate apparatus indicated are ones who would have been killed long before they reached their destination. But it is possible to conceive of an observer who did not have this physical limitation and so could have made a report at 4.00 p.m. on 1 January 1966. The fact that this sort of evidence, if multiplied, would always outweigh evidence from radiation received later on the surface of the Earth and similar evidence means that the statement really is a statement about the state of the Sun at 4.00 p.m. on 1 January 1966. The possibility of verification or falsification referred to is logical and not a physical one.

Now that this distinction has been made it should be clear that the interesting question discussed by philosophers about bringing about the past concerns statements really about the past. Could I, in a universe with different causal laws, bring about the truth of a statement really about the past. Could I really affect the past, e.g. make it not have rained yesterday?

Clearly I could not *change* the past — or the future either. To change something is to make it different from what it was at another temporal instant. I cannot make a thing at t_1 different from what it is, was, or will be at t_1 — for this would imply that it both did and did not have some property at t_1. But the question remains whether I could bring about what a thing was at a past instant by my present action.[1]

A recent claim that it is logically possible that we could affect the past comes in an article by Michael Dummett [9]. As his discussion brings out the issues more fully than other discussions such as those mentioned in the footnote to this page, I will consider his treatment of the question. In so doing I hope to show conclusively that it is not logically possible that anyone should be able to bring about a past state (or event).

Dummett describes the following as a situation in a possible universe and then suggests that we should describe it as a situation where someone affects by his present action what is past and gone:

> Suppose we come across a tribe who have the following custom. Every second year, the young men of the tribe are sent, as part of their initiation ritual, on a lion hunt: they have to prove their manhood. They travel for two days, hunt lions for two days and spend two days on the return journey; observers go with them, and report to the chief on their return whether the young men acquitted themselves with bravery or not . . . while the young men are away from the village the chief performs ceremonies — dances, let us say — intended to cause the young men to act bravely. We notice that he continues to perform these dances for the whole six days that the party is away, that is to say, for two days during which the events that the dancing is supposed to influence have already taken place ([9], pp. 348f).

Is it logically possible that the chief's dancing of the last two days could affect whether the young men had been brave? Suppose there is a strong correlation between bravery of the young men and the chief dancing in the last two days. We should naturally say in that case that the bravery was the cause of the dancing and not vice versa. But suppose that — to all appearances — it is in the chief's power to dance or not. Whenever the chief does dance, the evidence shows that the young men have been brave. If the chief dances after the observers' report that they had not been brave, the observers subsequently admit to 'lying'. ([9], p. 354). Dummett concludes that in this situation we should say that the chief's

[1] Both Ayer [2] and Russell [1] seem to have claimed that it is logically possible that men should affect the past. Both have urged that it is only the fact that we know so much about the past that makes us think that we cannot affect it. But, they claim, our relatively greater knowledge of the past than of the future is a contingent matter.

dancing subsequent to the hunt produced the bravery of the young men, made the young men have been brave. He says that normally we assume that a man can know whether a past event occurred independently of his present intentions. This sort of case, he argues, would, however, force us to reconsider this assumption.

There are here two issues which must be kept sharply distinct. One is the main issue of whether it is logically possible that an action such as the chief's dancing could have an effect at an earlier time. The other is whether any one could ever have any evidence that an action such as the chief's dancing had an effect at an earlier time. A verificationist (see p. 6) will hold that a negative answer to the latter question will entail a negative answer to the former question; for him claims are only logically possible factual claims if there can be evidence for or against them. I do not wish to take the verificationist line, and so I will produce one argument to show that there could never be any evidence that an action had a past effect (without begging the question whether it is logically possible that it could), and a second argument to show that it is not logically possible that an action have a past effect. Someone who was not convinced by the latter argument might still accept the former one.

Our grounds for believing that a past action has a future effect is that before the action our evidence suggests that one thing is going to happen in future, and after the action our evidence suggests that a different thing is going to happen. Analogously any defender of the claim that there could be evidence that an action had had a past effect must claim, as Dummett does, that the action must produce changes in the evidence about what happened in the past. Otherwise there would be no ground for supposing that the action had any effect in the past at all. Now I shall urge that no change in the evidence could possibly substantiate the claim that some one had affected the past.

The evidence about what happened need not take the form solely or at all of memory-claims by observers of the past event. It may alternatively take the form of traces of the past event. Some present state of the world is a trace of a former state (or event) if well-substantiated scientific laws show that such a state (or event) normally precedes such a state. The change in the evidence may therefore consist either, as in Dummett's example, of a change in the memory claims, or of a change in the traces.

Suppose then that the advocate of the possibility of affecting the past claims that an action at t_2, an instant subsequent to t_1, had the result that the evidence about what happened at t_1 (that is, the memory claims and observed traces about what happened) was different subsequently to t_2, let us say at t_3, from what it was at t_1 and up to t_2. The evidence at t_1 (and up to t_2) and the evidence at t_3 (and at all instants since t_2) suggest different occurrences at t_1. Then there are two alternatives. Either we realise at t_3 that a change has taken place at t_2 in the evidence or we do not. In the latter case we could never have any grounds for believing that we had affected the past, for we would have no knowledge of any change in the evidence as to what had happened. So there would be no place in our beliefs about the world for a belief that we could affect the past. In the former case we realise at t_3 that we did something at t_2 as a result of which the observed evidence at t_3 about what happened at t_1 is different from what it was at t_1, one or other evidence having been affected by the action. This suggestion we must now examine more fully.

Which is the better evidence about what happened at t_1 — the evidence observed at t_1 or that observed at t_3? The first possibility (one which Dummett does not consider) is that the evidence at t_1 might be better and the evidence at t_3 worse. But this claim cannot be made by an observer at t_3. For in claiming that the evidence at t_1 indicated more truly than that at t_3 what happened at t_1, he is claiming that the evidence available to him shows that the evidence at t_1 shows better what happened at t_1 than does the evidence at t_3. But the evidence available to the observer is the evidence at t_3. And what that shows is that what happened at t_1 is what was indicated by the evidence at t_3. Hence the suggestion that the evidence at t_1 might be the better evidence proves self-defeating.

The other possibility (the one which Dummett does consider) is that the evidence at t_3 might be the better evidence. This could arise for the reason that the evidence observed at t_1 was not typical (viz. had more evidence been observed at t_1, different conclusions could have been drawn). But in that case the past has not been affected by the action at t_2. The only effect of the latter is to improve the condition and appreciation of the evidence. But if it be said that the evidence observed at t_3 is better evidence despite the fact that the evidence observed at t_1 was typical of the total possible evidence, then the evidence at t_1 and at t_3 cannot be really about what happened at t_1. For, we concluded earlier, evidence about what happened at t_1 could always — it is logically possible — be obtained at t_1 better than at any other instant. If the evidence at t_3 is more to be relied on than any amount of evidence at t_1, then what it is evidence of cannot really be what happened at t_1: the description allegedly of what happened at t_1 must have a covert future reference.

Let us apply these points to Dummett's example. The young men return and the observers report that they have not been brave. The chief then dances. The observers then say that they previously gave untrue reports. The young men really had been brave after all. Now the observers may believe that they lied, that they knew all along that the young men had been brave. In which case they would have no grounds for saying that the chief's dance had affected the past; the only effect of the chief's dance was, they must conclude, to make them now tell the truth. If others who heard the two reports believed that the observers had lied, they too must conclude that the only effect of the chief's dance was to elicit truth from lying observers. On the other hand the observers might now believe that the young men had been brave without believing that they had believed this when they gave their first report. They would not in that case, if they were now being honest, say that they had lied previously, but that they had been mistaken. Despite Dummett's choice of the word 'lying' to describe what happened, a defender of the possibility of affecting the past might describe the situation in these terms. Now the observers or anyone else who accepted this description of what happened might be making one of two possible claims. The first is that at the instant of the occurrence, the observers were deceived. The evidence at the instant of the occurrence was that the young men had been brave, but that evidence was not fully observed or was insufficiently appreciated at that instant. The observed evidence was untypical of the total possible evidence and hence the first report of the observers was mistaken. So the effect of the dance is to give a correct estimation of the previous experience. It enabled the observers to see that they had been subject to a deception. The dance affects the subsequent state of mind of the observers. There is again no question of it affecting the past.

But the observers' assertion might be interpreted in a different way. When the observers claimed that they were mistaken in their previous report, they might be claiming that while the first report was based properly on the available evidence and that the available evidence was a representative sample of the total (logically) possible evidence; nevertheless they now judged on subsequent evidence that the young men had been brave, and that this latter evidence was to be preferred. But in this case the statement 'the young men were brave' is not a statement solely about what happened on the lion hunt, but, at any rate partly, about something subsequent, e.g. about what the observers were naturally inclined to claim about what happened on the lion hunt. The earlier distinction in this case delimits 'the young men were brave' as a statement with a covert future reference. But in that case the fact that the chief's action affects its truth value is no more puzzling than the fact that future occurrences will affect the truth-value of 'Father Jones baptised in 1964 the next century's greatest pianist'. For in both cases the influence of the affecting action is really in the future. I conclude that on none of the alternative interpretations do the observers or any one else have reason to suppose that the chief's dance affects what is past and gone. And generally no one can have any reason to suppose that an action at t_2 can affect what happens at an earlier time t_1.

Note that the above argument has no tendency to yield the absurd consequence that we cannot have reason to believe that an action at t_2 cannot have an effect at t_3. For after t_3 we may have just such reason. We may know that there was a change at t_2 in what the evidence suggests is going to happen at t_3. We take the evidence at, and around, t_3 as showing what in fact happened at t_3 and therefore as better evidence than the evidence at t_1. The discrepancy shows that what happened at t_2 made something to happen at t_3 which would not otherwise happen.

But even if we would never have reason to believe that an action C at t_2 brought about an event E at an earlier time t_1, is it not logically possible that it should bring it about all the same? I do not think so. For if C brings about E, then the performance of C must make the difference to whether or not E occurs. But it is logically possible that any event such as C should have no cause, that an agent should freely choose at t_2 to bring about C without any cause making him to choose. In that case it could not be predetermined before t_2 that E occurs yet some one could observe E happen at t_1 and know as surely as could be known that E was happening then; and still it would be not yet predetermined whether or not E would happen; it could be that a man's future choice might make E happen or alternatively make it not happen. Here there does seem to be a contradiction. It cannot be the case both that E is now happening and that it is yet to be determined whether E will happen. And so it is not logically possible that a man should know the former if the latter is true. It is therefore not logically possible that a man could affect a past state by his present action. Actions can only have present and future effects — and they could only have present effects if influences could be propagated with infinite velocity.

Now granted that an agent cannot by his action affect what is past and gone but only what is to come, it follows that no state (or event) can have its cause subsequent to it.[1] For if we say that in certain circumstances an event A was the cause of a state B, we thereby commit ourselves to the proposition that a man could have made B happen by making A happen.

---

[1] Several recent philosophical articles have made this point, producing different arguments in support of it. See [4], [5], [7] and [11]. See [3],[9] and [12] for the opposite point of view.

But since, as we have seen, it is not logically possible that a man can make *B* happen by making *A* happen if *B* is prior to *A*, then *A* cannot be the cause of *B* if *B* is prior to *A*.

One might attempt to avoid this conclusion by denying that the implication of the last paragraph holds in cases where *A* is outside human powers of control. One might thus claim that to say that in certain circumstances *A* is the cause of *B* does not commit you in cases where *A* is a kind of state (or event) outside human control to saying that if a man were to make *A* happen in those circumstances he would thereby bring about the occurrence of *B*. One would thus be claiming that what is involved in saying that *A* is the cause of *B* is different in cases where *A* is something lying outside human control. Thus 'cause' would have a different meaning in 'The explosion of the bomb was the cause of the devastation' from 'The explosion of a star was the cause of the gas cloud', because we can make bombs but not stars explode. But this is not plausible, for the boundaries of what humans can do are continually extending without the meaning of 'cause' thereby changing. 'Chromosomal mutations cause the appearance of new phenotypic characteristics in organisms' did not change its meaning when men became able to produce chromosomal mutations. There is no reason to suppose that the meaning of 'The explosion of a star was the cause of the cloud' will change when humans learn to make stars explode. Rather 'cause' retains its meaning, while the boundaries of what humans can cause change. I conclude that of logical necessity no cause can follow its effect and hence that of logical necessity the past is determinate.

States of objects at any instant t_2 of logical necessity cannot, we have shown, be affected by states subsequent to t_2. Hence either they are uncaused (they just happen, nothing produces them) or they are produced by states at t_2 or prior to t_2. Now consider the state of the Universe in a region R_1 at a time t_1, which we will call state S_1, and the state of the Universe in that or some other region R_2 at a later time t_2, which we will call S_2. Now, as we have seen, it is not logically possible that S_2 be the cause of S_1. But it is always logically possible that S_1 be the cause of S_2. We can conceive of evidence from other states at other periods of time which would confirm the theory that S_1 was the cause of S_2. The evidence would be observations that states similar to S_1 in regions similar to R_1 are always followed by states similar to S_2 in regions similar to R_2, the regions having to each other the relation that R_1 had to R_2 and the states occurring at the time interval $t_2 - t_1$; or the evidence could be more indirect evidence showing that such connections were to be expected in nature. Whether or not S_1 is the cause of S_2 is a logically contingent matter and depends on what are the laws of nature governing the succession of states. This being so, it is always logically possible that any specified future state of a region of the Universe be caused by any specified present state; whereas, as we have seen, it is not logically possible that any past state of a region of the Universe could be caused by a present state.

From these basic principles a number of important necessary truths about time will be proved in this and subsequent chapters. The first of these which I shall demonstrate here is that the present instant cannot ever return, that is that time cannot ever be closed. If the present instant t_1 will return, then the next instant subsequent to this one, t_2, will be both before and after t_1. Yet in virtue of being before t_2 the state of any region of the Universe at t_1 is of logical necessity unaffectable by any state at t_2. Whereas in virtue of its being after t_2, it is logically possible that any state at t_1 could be affected by the state at t_2. Hence the supposition of a cyclical or reversing time leads to self-contradiction.[1] This is not to deny that it makes sense to talk of a cyclical or reversing Universe. To say that the Universe is cyclical is to say that after so many years its state is exactly similar to what it was and an exactly similar series of events take place again. S_1 is followed by S_2, S_2 by S_3, S_3 by S_1, S_1 by S_2 and so on *ad infinitum*. The Stoics believed in just such a cyclical Universe. But the point is that S_1 comes again at a later temporal instant, not at the same instant. So with a reversing Universe. To say that the Universe is reversing is to say that after so many years an exactly similar series of states of the Universe to the series which has occurred

---

[1] This point suffices to rule out on purely logical grounds one cosmological model of the Universe developed from the General Theory of Relativity, the Gödel model. The Gödel model has closed time-lines allowed the physical possibility of an agent influencing the past. Gödel claimed that it would be, at any rate in 1949, practically impossible for an agent to achieve a sufficiently high velocity relative to his galactic cluster $\left(\text{at least } \dfrac{1}{\sqrt{2}} \times \text{the speed of light}\right)$ to have effects in the past, and seemes to think that this avoids the logical difficulties of his model. But if his model allows as physically possible something which cannot be physically possible because it is logically impossible, the model must be wrong. For the mathematical presentation of his model see K. Gödel, 'An Example of a New Type of Cosmological Solutions of Einstein's Field Equations of Gravitation' in *Reviews of Modern Physics*, 1949, 21, 447–50. For his attempt to deal with the logical difficulties of the model see K. Gödel, 'A Remark about the Relationship Between Relativity Theory and Idealistic Philosophy' in *Albert Einstein: Philosopher-Scientist* (ed. P. A. Schilpp) New York, 1949 and 1951, pp. 555–62. See especially pp. 560f.

occurs in the reverse order to that of its original occurrence. S_1 is followed by S_2, S_2 by S_3, S_3 by S_2, S_2 by S_1. But the point, as before, is that S_1 comes again at a later instant, not at the same instant[1].

BIBLIOGRAPHY

For discussion of the priority of cause to effect, See:
[1] B. Russell, *Mysticism and Logic*, London, 1918, pp. 201ff.
[2] A. J. Ayer, *The Problem of Knowledge*, London, 1956, pp. 164–75.
[3] Michael Dummett, 'Can an Effect precede its cause?' *Proceedings of the Aristotelian Society*, Supplementary Volume, 1954, 28, 27–44.

[1] In [8] Grunbaum appeals to Leibniz's principle of the identity of indiscernibles to establish the logical possibility of cyclical time. He describes a universe in which an exactly similar series of states to any given series has occurred before and will occur after it again and again *ad infinitum*. This is clearly, as I claim in the text, a logically possible Universe. He then urges that this Universe ought to be described, not, as I have described it, as a cyclical Universe, but as a Universe in a cyclical time. For, he urges, following Leibniz, 'if two states of the world have precisely the same attributes, then we are not confronted by distinct states at different times but merely by two different names for the same state at one time.' Leibniz's principle from which this particular argument is taken is that any two things which have all the same properties, including relational properties, are really the same thing. Each occurrence of S_1, Grunbaum claims, has such properties as being after a state S_3, and being before a state S_2 which is in turn before a state S_3, and so each occurrence is the same occurrence.

However Leibniz's principle may be interpreted in a weak or a strong sense, according to which properties we take into account. In the strong sense we take into account only properties picked out by predicates which do not mention named objects — e.g. being two metres from an obelisk or being in a man's right hand. In the weak sense we take into account all properties including properties picked out by predicates which mention named objects — e.g. being two metres from Cleopatra's needle, or being in my right hand. But if we interpret Leibniz's principle in the strong sense, there are good arguments against it. For example could there not be a universe consisting only of two qualitatively identical steel balls? Each would have the same properties as the other (e.g. being round and being two metres from a round steel ball), and yet there would be two of them. Yet if Leibniz's principle is understood in the plausible weak sense, a cyclical universe does not violate it. For the series of exactly similar states . . . S_1, S_2, S_3, S_1, S_2 . . . etc., would consist of a series of namable states $S_1{}^a, S_2{}^a, S_3{}^a, S_1{}^b$, $S_2{}^b$ . . . etc. $S_1{}^a$ would have the properties of being identical with $S_1{}^a$, occurring immediately before $S_2{}^a$, etc; and these would be properties which other S_1s lacked. And so on. On these points, see for example, Max Black, 'The Identity of Indiscernibles', in his *Problems of Analysis*, London, 1954.

[4] Antony Flew, 'Can an Effect precede its cause?' *Proceedings of the Aristotelian Society*, Supplementary Volume, 1954, 28, 45–62.

[5] Max Black, 'Why Cannot an Effect precede its cause?' *Analysis*, 1956, 16, 49–58.

[6] Roderick M. Chisholm and Richard Taylor, 'Making things to have happened', *Analysis*, 1960, 20, 73–8.

[7] William Dray, 'Taylor and Chisholm on Making things to have happened', *Analysis*, 1960, 20, 79–82.

[8] Adolf Grünbaum, *Philosophical Problems of Space and Time*, London, 1964, or second enlarged edition, Dordrecht, Holland, 1973, pp. 197–203.

[9] Michael Dummett, 'Bringing About the Past', *Philosophical Review*, 1964, 73, 338–59.

[10] Samuel Gorovitz, 'Leaving the Past Alone', *Philosophical Review*, 1964, 73, 360–71.

[11] Richard M. Gale, 'Why a Cause cannot be later than its Effect', *Review of Metaphysics*, 1965, 209, 209–34.

[12] J. L. Mackie, *The Cement of the Universe*, Oxford, 1974, especially ch. 7.

On the continuity of time see:

[13] J. R. Lucas, *A Treatise on Time and Space*, London, 1973, § 6.

9 Logical Limits to Spatio-Temporal Knowledge

I wish in this chapter to consider how we acquire knowledge of events at other places and temporal instants, and, in so doing, to show that there are certain logical limits to the amount of knowledge which we can have about events at other temporal instants and to the ways in which we can acquire it.

There are two ways by which we may come to know about objects, their states and changes of state at other places and temporal instants than the present place and temporal instant.

First we may be in a position to give reports of what we have observed or will observe at other instants. We do not in this case infer what happened at those other instants from something else, for we can report straight off what happened or will happen, and in what order the events happened or will happen. Claims to knowledge of past events as a result of having observed them are claims to memory, and these we often make. I can say without needing to infer it from anything that yesterday we had steak for dinner and the day before we had fish and now we are having chicken. Claims to knowledge of future events on the basis of being about to observe them would be claims to foreknowledge. Claims to foreknowledge are of course seldom made. We claim non-inferential knowledge of the past but not of the future. Any claims to memory, as to foreknowledge, may of course be mistaken, and for what I say without inference to constitute knowledge of what happened it must not be denied by other people; and other people must corroborate some of my reports, in order that others of my reports which are neither denied nor confirmed by other people may be taken as evidence of what happened. If many others disagree with me and hold a common view about what happened, it is generally agreed that my claim to knowledge is not justified. It may seem logically possible that we might have corroborated foreknowledge of what we will observe similar to the corroborated memory of what we have observed, which we do have. I shall discuss how far this is a logical possibility later in the chapter.

There is one qualification which must be added to the remarks in the last paragraph. Only in the case of objects and their states spatially near to ourselves will the instant at which we observe the objects be (to any very high degree of approximation) the instant at which the objects are in the states observed. We know that the instant is virtually the same because we observe a spatially near change which we bring about at virtually the same instant as we bring it about. I know the present instant at which I am observing a certain state of the gum pot on the table is approximately the instant at which the gum pot is in that state, because if I knock the pot over, I observe it knocked over as soon as I have performed the action. This applies to visual and almost any other kind of observation. On the other hand, any signal including light sent to a distant object and reflected thence takes some period of time from when I send it to when I observe it again, and thus I conclude that I observe distant objects as they are at more remote temporal instants than when I observe them. Hence, if a man reports that first he observed a state A_1 and after it a state B_1, he is only producing evidence that A_1 was followed by B_1 if A_1 is spatially near. Otherwise further investigation is needed to ascertain which came first.

Secondly we have what I shall term traces of what has happened and of what will happen at our present place or other places, and also of what is happening now at other places. By a trace I mean an event or some state of an object which is clear evidence for an observer with his knowledge of how things work in the world that some other event or state very probably has occurred, is occurring, or will occur. 'Trace' is a technical term introduced by me. In normal parlance it may be the case that traces are only of what was and not of what is or will be, but in my use of the term traces can be equally of what was, what is, and what will be. 'Trace' is a subjective matter, in the sense that what is for one man a trace of some event may not be one for another whose knowledge of how things work in the world is different. A detective may spot immediately from the peculiar state of a broken lock that Fagan broke it; most men would not recognise the state of the lock as indication of Fagan's activity. The state of the lock is therefore, I will say, a trace for the detective but not for most men of Fagan having broken the lock. However, as most humans have largely similar knowledge of how the world works to each other, what is a trace for one is a trace for another, and so I shall be able to ignore personal idiosyncrasies in talking about traces. The traces which I shall discuss will be traces for all men or at any rate for all men of a certain culture. Examples of traces (which are traces for most Englishmen anyway) are the following. Footprints in the sand are a trace that someone has walked on the sand recently; a track on a tape-recorder is a trace that someone has recently sung the song recorded on it. A fall in the barometer is a trace of impending rain. An announcement in *The Times* that a certain meeting is taking place today is a trace that it is indeed taking place.

I must make here a distinction which will be of use later between what I have termed traces and what I shall term signs. A_1, a particular state or event of some type A, is a sign of B_1, a particular state or event of some type B, if the occurrence of an A is highly correlated with the occurrence of a B. If the correlations are between precedent As and subsequent Bs, then A_1 is a sign of a future state or event B_1 and so on. 'Sign' like 'trace' is a technical term introduced by me. Signs are an objective matter. An event being a sign of some other event is a relationship between kinds of states and events in nature, whether or not men know about it. Hence to say that some event is a trace for me of some other event is to say that on the evidence available to me I believe with reason that it is a sign of the other event.

Traces and reports of observers must not conflict if they are to be taken without further investigation as reliable evidence of what happened, is happening or will happen. If the traces and reports of observers do conflict, more thorough investigation is needed before we can say what happened, is happening, or will happen. The trace then is taken as reliable evidence only if its connection with that which it signifies is well-established and apparently invariable. Reports of observers in these circumstances are only taken as showing what happened if there are many corroborated reports. Traces and reports are weighed against each other. Thus if a night-watchman claims that nobody entered the building during the night, but there are wet muddy footprints in the building, the night-watchman's report is suspect. If Jones' fingerprints are found near the scene of a crime, yet many otherwise reliable witnesses report that he was far away from the scene of the crime during the period when it was committed, it becomes doubtful whether the fingerprints are a reliable trace of Jones's presence. Perhaps they were planted. The reliability of a particular trace may be impugned by reports of observers, and particular reports of observers by traces. The stronger evidence wins, especially if a plausible account can be provided of why the weaker evidence is on this occasion misleading.

Yet while in judging what happened on a particular occasion traces and reports of observers are weighed against each other and support or impugn each other, in an important sense reports of observers provide the basic evidence of what happens at other places and temporal instants, since traces depend ultimately for their reliability on reports of observers.

Our most usual grounds for taking some particular state or event as a trace of some other state or event is that observers have reported that states or events similar to the first are normally accompanied by states or events similar to the second. I take a present state A_1 as a trace of a past state B_1 because observers, including possibly myself, report that whenever an A occurs, a B has always or almost always preceded it (I use the capital letters without subscript for a type of event or state, and the letter with subscript for an individual event or state of the type referred to by the capital letter). The footprint is a trace that someone has walked on the sand recently because in the experience of men footprints (in the sense of marks in the shape of human feet) are normally preceded by someone walking on the sand. We take a present state C_1 as a trace of a future state D_1 because observers report that at other times whenever a C occurs, a D always or almost always follows it. The fall in the barometer is a trace of impending rain because in the experience of men falls of the barometer are normally followed by rain. The announcement in *The Times* is a sign that the meeting is now taking place because men have experienced that announcements in *The Times* of events for the day of publication are normally correlated with the occurrence of those events. In so far as the reports are many and corroborated, and the reports are of invariable correlations (e.g. Cs being followed by Ds without exception), by so much the trace is better evidence of what it indicates. We may use more complicated combinations of evidence of the kind described. We may take an event F_1 as a sign of a precedent event H_1 by using a chain of evidence. We may never have observed Fs being preceded by Hs, but we may have observed that Fs are normally preceded by Gs, and Gs by Hs. It is in this way that we conclude that some present geological state is evidence of a long-past state.

Our grounds for taking some event as a trace of another may be more remote. Various miscellaneous observed connections between events may be most simply explained by setting up some scientific theory, a deductive consequence of which is the law 'All As produce Bs'. This law then allows us to take an A_1 as a trace of a future B_1. No one may ever have observed any A being followed by a B, but the evidence for the truth of the law may be more indirect. Thus by studying stellar spectra and masses, we may obtain evidence for a theory of stellar evolution. One law derivable from this theory might be the law that stars of a certain type explode after reaching a certain stage of evolution, viz. the law that a certain state of a star A produces a subsequent event of an explosion B. We establish the law without ever having observed stars of the type in question exploding. Hence when we observed a particular star in a state A_1 we would take this as a trace of a future explosion B_1. Yet by whichever way we learn to take A_1 as a trace of B_1 our justification for doing so lies in the reliability of the reports of observers. For this reason reports of observers provide the ultimate evidence of what happens at other places and temporal instants. The procedure of postulating that observed connections or inferred laws operate beyond the range over which they have been observed to operate is the procedure which I have called in earlier chapters extrapolation.

There is an important asymmetry between the use of scientific laws to establish traces of future events and their use to establish traces of past events, between prediction and retrodiction. Scientific laws state how events and states of one type, As, cause events or states of another type, Bs. They may not use the word 'cause', but this is their form. They state the consequences of the occurrence of an A, however the A may be brought about. Consequently they are of no use as they stand for retrodiction, without a further premiss. We may have good evidence that As cause Bs but that by itself does not allow us to take B_1 as a trace of a precedent A_1. For many other states than As may cause Bs. Only if we have grounds for believing that B_1 had a cause and that any other cause of Bs than As would have produced other effects which As do not produce and these are not observed, can we take B_1 as a trace of precedent A_1. Our grounds for supposing that B_1 had a cause is that science has shown us events usually do have causes, and explanations can be provided for the previously mysterious. But we would have reason to believe that B_1 did not have a cause if we could show that only a certain number of states could bring about Bs and that none of them were operative on this occasion. How we could show this will be illustrated at some length in Chapter 15 when we shall be discussing an important example of a state which, it is suggested, had no cause. If then we suppose that B_1 has a cause, in order to show that it was A_1 we have to show that any other cause of Bs than As would have produced other effects as well, but that these have not occurred. But we can only investigate the possible occurrence of effects of a finite number of possible causes. So only if we have reason to believe that only one of a small finite number of possible causes was responsible for B_1 can we test to show that A_1 is responsible. But what reason could we have for supposing that only one of a certain small number of possible states C_1, D_1, E_1, F_1, etc. could have brought about B_1? One reason we could have is that we (or other observers) have observed the state of the Universe in the relevant region at the instant of the purported occurrence of A_1. But then if we or they had done that we would not have needed to use retrodiction to ascertain what happened then. The occurrence of B_1 would merely confirm reports of observers that A_1 had occurred. But suppose that we have no observers' reports, that we are using retrodiction as an independent source of information about the Universe. In that case our grounds for supposing that no cause other than one of a certain small finite number of states can have brought about B_1 can only be that other causes of Bs have seldom or never been observed to occur. But this is not, of course, conclusive evidence that no such cause did occur on the occasion in question. Hence even when we have established the laws of nature, there is an essential weakness in retrodiction which there is not in prediction. Let us take an example. A skull of a certain type with its carbon atoms in a certain state in a certain geological deposit is a trace that men lived in the region a million years ago. We take the skull as a trace of human habitation because:

1. We believe that human heads turn into skulls in this condition after a million years. This is a well, albeit indirectly, established scientific law. Hence we have a possible cause.
2. A few other known possible causes of the skull being in that condition have left no other effects; viz. there is no evidence of forgery of a known type.
3. We believe that other possible causes of the skull in that condition are seldom if ever operative.

But our only grounds for the latter belief is that we have never observed them in operation. These are strong grounds, but they do not show at all conclusively that another cause could not have been operative in this instance. On this occasion perhaps an unknown cause, viz. a cause at the description of which we cannot guess, may have brought about B_1. We cannot conduct any experimental test to refute this suggestion — for reasons of logic.

If someone suggests that an unknown effect is caused by some event H_1 in certain circumstances, we can — it is logically, if not always practically possible — do an experiment to confirm or refute the suggestion. We could produce an H in similar circumstances and see what happened. But for reasons of logic, as we saw in the last chapter, we cannot do a comparable experiment to test whether an unknown cause brought about a certain effect. For we cannot make things to have happened. Hence there is of logical necessity a weakness involved in retrodiction which is not involved in prediction. Scientists do not postulate backward-moving laws because they could never be experimentally tested, that is, tested in the way just described.

Now it is true that some forward-moving scientific laws cannot be experimentally tested, and this is true in particular of cosmological laws, laws predicting a future state of the universe millions of years ahead caused by a present state. This cannot be tested experimentally because humans cannot bring about states of the Universe. But there are two differences from the case of any purported backward-moving laws. First the limit is physical or practical, not logical. It is just a contingent matter of fact that we cannot bring about a certain state of the Universe and see what happens. Secondly, cosmological theories can be subjected to indirect experimental test, because a law forming part of a more general theory, viz. normally the General Theory of Relativity, from which they are deductively derived by the addition of astronomical data, can be subjected to experimental test on the small scale. But no purported backward-moving law could be subjected even to indirect experimental test, for the only theoretical law of which it would be a deductive consequence would be a backward-moving law and this again — for reasons of logic — could not be subjected to experimental test.

The use of scientific laws to establish that a state is a trace of some other state at the same temporal instant, suffers from the same weakness as the use of scientific laws for retrodiction and for the same reason — given, as scientific evidence shows, that no effects are propagated with infinite velocity. For given this, no law stating that one state is correlated with a contemporaneous one could be experimentally tested. Hence science does not purport to put forward any real laws of co-presence, any more than it puts forward backward-moving laws. Certainly there are a number of well-established scientific principles, stating how various contemporaneous characteristics are correlated. Biology is full of such principles as that tortoise-shell cats must be female. But the form of such principles is misleading. The zoologist in stating that tortoise-shell cats must be female does not really mean that of physical necessity if a genetic engineer makes a cat to be tortoise-shell, he thereby causes it to be female. What he means is that tortoise-shell cats bred from the current stock of the cat species by normal processes must be female. But in that case the principle is not a law of co-presence. The principle states that whatever arrangement of a genotype drawn from the genetic pool currently available to the feline species produces cats of tortoise-shell colour also produces the normal female characteristics in them. The principle is in fact a summary of two forward-moving laws and a historical statement about the genetic pool currently available to the feline species.

Any proposed real law of co-presence would be unable to be tested experimentally, just as any proposed backward-moving law would be unable to be tested experimentally. For suppose a scientist claims as a law of nature that a state A is always co-present with a state B. Now given that effects are never propagated with infinite velocity, the state A cannot be the cause of the state B or conversely. Hence if the one state is correlated invariably with the other, it must be because the two states have a common cause. But then one could only prove the correlation to be invariable if one could prove that it was physically necessary that there be a common cause. But for the reasons given above one cannot prove at all conclusively that some effect must have a certain cause, and so one cannot prove at all conclusively that two characteristics must be correlated. Genetic engineers of the next century may well produce male tortoise-shell cats.

So then we may use traces to infer the occurrence of past, present, or future states. We may take a state as a trace of another on the basis of directly observed connections or on the basis of an inferred scientific law. If the scientific law established is true and concerns invariable correlations (and includes all the necessary qualifications for surrounding circumstances) then the trace of the future state which is predicted by the law will be absolutely reliable. Traces of both past and present states could not be proved to be such by experimental test — for reasons of logic; and hence there are logical limits to the kind of confirmation which they could receive of their status. Proof that a trace is a trace of a past or present state depends at a crucial point on mere observational correlations, about which it cannot be tested experimentally whether they hold universally.

Evidence from one trace may conflict with evidence from another trace. We may have good reason to take an observed present state A_1 as evidence of another present state B_1 and also good reason to take an observed present state C_1 as evidence of another present state D_1 which could not co-exist with B_1. Here of course we take the best substantiated connection as providing the most reliable indication. If the connection of As and Bs is better established than the connection of Cs and Ds, then the two observed states A_1 and C_1 taken together are traces of a present state B_1. They show that the connection between Cs and Ds does not hold universally. We saw this process at work in Chapter 5 (pp. 94ff) where estimates of the distance of distant objects which assumed one regular connection to hold universally were opposed to other estimates which assumed a different connection to hold universally. The issue turned on which connection was best established. One which was a consequence of a well-established theory of mechanics was better established than one which depended on mere observation of a few instances.

The supposition that observed connections hold beyond the range for which they have been tested becomes less and less reliable, the greater the difference between the range studied and the range in which they are supposed to hold. We have reason to assume that what is a trace within the observed region is a trace in a region outside that observed — so long as the unobserved region does not have characteristics, including spatio-temporal location, too radically different from the observed one. Suppose that I was born at a place P in the Sahara desert and that my knowledge of the Earth is confined to a region of radius fifty miles surrounding P. This I have explored, but I have no knowledge by personal exploration or from reports of other travellers of regions more distant. I thus conclude that a square mile of sandy desert at some distance from P is an unfailing trace that there is another square mile of sandy desert at a distance from P greater by one mile. Clearly I am justified in using these traces to reach conclusions about the geography of regions up to, say, sixty or seventy miles from P, that there is nothing but sand there. I reach such conclusions by taking an observed square mile of sand as a trace of a square mile of sand beyond it and the latter, to which I have inferred, as a trace of a square mile of sand beyond it, and so on. But clearly the further away I claim to have knowledge of the geography of the Earth, the less justified are my claims. Claims to knowledge of the geography of regions 1000 miles from P on the basis of an examination of regions within fifty miles of P would clearly be very weakly justified. The further from P I stake my claim to knowledge, relatively to the size of the region which I have examined, the less justified my claim will be. This is because the region to the properties of which I infer will differ more and more, at any rate in respect of spatial location, from the region which I have studied. The possibility looms larger and larger that the region which I have examined is untypical of the regions of the Earth about which I am making inferences. Traces claimed to be reliable on the basis of examining a small region become less and less reliable, the further they are extrapolated. Similar arguments apply to the reliability of inferences to events at different temporal instants.

The observations which we make in order to reach conclusions about a temporal succession (though not about co-presence) must be observations of temporal successions. Only on the basis of directly perceived temporal successions can we infer to others. Some writers on the philosophy of science have, however, wanted to urge that knowing that certain events happened we could arrange those events in temporal order without basing that arrangement on any direct perception of temporal succession, but simply by means of the principle enunciated in the last chapter that causes precede their effects. Now certainly of any two particular events A_1 and B_1, if we can say that B_1 is cause and A_1 effect, then we can say that B_1 happened before A_1. But how can we know that B_1 is cause and A_1 effect unless we have observed that in similar circumstances As normally follow Bs, but Bs do not normally follow As, or had more complicated evidence of temporal succession, such as that Gs follow Hs, but Hs do not follow Gs, from which we constructed a theory, a deductive consequence of which is that As cause Bs? If all we knew was that As and Bs or some other phenomena occurred we could suppose that they had occurred either in one order or in the reverse order. On one supposition we would conclude that A_1 must precede B_1, on the other that B_1, must precede A_1. We have to observe straight off many temporal successions in order to infer others.

All the writers who have attempted to derive conclusions about temporal order from premises not referring to it and thus set up what they have called a 'causal theory of time' are guilty of some *petitio principii*. The most celebrated example is Reichenbach. Reichenbach claimed that we could find out which of two events was cause and which effect by the following principle of temporal ordering: 'If E_1 is the cause of E_2, then a small variation (a mark) in E_1 is associated with a small variation in E_2, whereas small variations in E_2 are not associated with variations in E_1' ([1] p. 136). Thus suppose we have a beam of light passing between one slit (its presence there being the event E_1) and another (its presence there being the event E_2). We wish to ascertain which way the light is streaming, viz. whether E_1 is the cause of E_2 or vice versa. We therefore put a piece of green glass in the way of the light at one instant at the first slit (viz. modify E_1 by making the light at the first slit green) and then at another instant at the second slit (viz. modify E_2 by making the light at the second slit green). In this way, Reichenbach claimed, we could discover which of E_1 and E_2 were cause and effect without relying on any direct, viz. non-inferred, perception of temporal order. For, indicating marked events (that is light made green) with an asterisk: 'We observe only the combinations $E_1 E_2$, $E_1^* E_2^*$, $E_1 E_2^*$ and never the combination $E_1^* E_2$' ([1], p. 137). Hence we conclude that E_1 is the cause; because if it is affected, E_2 is also — but the converse does not occur. The fallacy should be apparent.[1] How are the pairs grouped? The events E_1, $E_1{}^*$, E_2 and $E_2{}^*$ have occurred a number of times. So much we know without presupposing direct perception of temporal order. But in judging that $E_1{}^*$ went in a 'combination' with $E_2{}^*$ rather than with E_2 which also occurred, we are relying on the fact that we perceived $E_2{}^*$ to occur immediately before or after $E_1{}^*$ whereas we observed E_2 to occur on another occasion.

So then some judgements of temporal precedençe must be made in virtue of direct perception of temporal order by observers. However, once we have made some judgements of temporal priority from direct observation, we can make other judgements of temporal priority on the basis of the first judgements. Thus if I have noticed that in other cases craters of a certain type are always preceded by bomb explosions, but that the occurrence of such craters is not generally followed by bomb explosions, then, knowing that within a small temporal period there was a crater formed and also an explosion at a certain place, I can judge that the explosion preceded the formation of the crater.

It is, we have argued, by traces and reports that we learn about events at other temporal instants and places. If we learn about a past event or state of some object by observing a trace of it, where the trace is an effect produced by it, and the trace is one from which we become accustomed to make inference to the past event, than we may be said, in a derivative sense of the term, to observe the past event or state.

An object or property or event is observed by an observer in the usual and obvious sense if he has knowledge of it as a result of stimuli from it impinging on his eye or other sense organ. An object or property or event is observed in a derivative sense if the observer observes in the usual sense effects produced by it which can be readily interpreted by the observer so that he knows about the object, property, or event. Thus we are said to observe the structure of the cell wall of the amoeba when we observe in the usual sense a photograph of it produced by an electron-microscope. Distant galaxies are said to be observed if we observe in the usual sense photographs of them produced with the aid of a radio-telescope. To say that something is observable is to say that either it is observable in the usual sense or it is observable in the derivative sense, by which is meant that either its effects can now be readily interpreted or that a machine by which the effects could be turned into traces which could be easily interpreted as effects of that thing could be constructed.

---

[1] It has been well demonstrated by, for example, Grunbaum [2].

We allow 'observation' by radio-telescopes, radar screens or television sets to be called observation because such instruments can be regarded as extensions of our sensory apparatus. Normally when we observe something, the thing produces effects on us as a result of which we have knowledge. But when we learn about the distant galaxies, they produce effects on an instrument which produces effects on us as a result of which we have knowledge. The instrument is only a finer detector than our sense organs. It is a contingent matter that our sense organs do not have its powers of discrimination. Hence we think of it as an extension to our sensory apparatus and term knowledge obtained by it knowledge obtained by observation.

But if an observer has knowledge of some event, his possession of which is not caused by the event, he is not said to observe the event. Even if he has a picture of the event before his eyes or in his 'mind's eye', he is still not said to observe the event unless in some way or other the occurrence of the event makes him know about it. Thus, suppose a parade is planned in meticulous detail by making in advance a detailed film showing how the soldiers are to march. The parade then takes place, and at the same time the film is shown to an audience somewhere else. However well the instructions were executed at the parade and however thoroughly the audience was convinced that they would be, the audience would still not be said to be observing the parade.

Now it follows from this that of logical necessity we can only observe what is present (and that in a narrow sense of 'present' only if effects are simultaneous with causes, viz. effects spread with infinite velocity) or what is past. My observing something involves it directly or indirectly producing an effect on me. By the principle established in Chapter 8, it is logically necessary that effects do not precede their causes. Hence we can have knowledge by observation only of what is present or past. We can predict the future but not observe it. Certainly one could construct an instrument on the screen of which could be produced a picture of the future state of some system. A machine, for example, might have parts which detect the present position of planets in the sky and it might work out their future positions and show them on a screen. But when we observe such a picture we cannot be said to be observing a future state. For such a machine could not — of logical necessity — be an extension to our sensory apparatus. What we are observing when we look at the picture is an effect of a cause (*not* an effect) of future effects. There is this vital contrast with the cases earlier considered which makes it natural to call those cases cases of observation, but not this case. It is a logical matter, not a contingent matter, that our sense organs could not be such that if the apparatus were part of them, we could be observing a future state.

Since we have used our knowledge of which states are traces of past states to extend our powers of observation, the correctness of those judgements must suffer from any inherent weakness in the kind of confirmation that can be had that a present state is a trace of a past state. As we have seen, there is a limit to the kind of evidence that can be provided for the reliability of such traces.

It is appropriate at this point to consider the logical possibility of time travel. Could I get into a machine, press a lever, visit past or future and return to the present? Here again the logical limits to the possible are provided by the principle that causes must not follow their effects. I cannot now cause anything to have happened. Hence I cannot get into a machine and press a lever which lands me at Hastings in 1066. For this would mean that by pressing a lever I would cause a machine to have been at Hastings in 1066. Now I could get into a machine and press a lever which landed me at Hastings at 2066 — the machine might squirt a drug at me so that I was deep-frozen until that year and the passing of time was unnoticed by me. But having got there I could not get into the machine and return to 1968, since I could not cause a prior event. The logical principle of causes not following their effects rules out this sort of travel — I cannot get to the past.

The only kind of time travel into the past that does make logical sense is that I should get into a machine, press a lever and see the past on a screen. Whenever we look at objects distant from the Earth we see them at temporal instants well past, for light takes a considerable finite time to travel the intervening space. The nearest star to ourselves apart from the Sun, Proxima Centauri, is seen as it was four years ago. There is no logical reason why I should not observe events at near-by places on Earth in the distant past. I have only to construct a machine which will interpret effects caused by the event and produce a picture of it. The difficulties of my observing the battle of Hastings on a screen are practical, not logical. But time-travel of a similar kind into the future is not logically possible. The difficulties of my observing now the next eclipse of the Sun on a screen are logical. I could get into a machine, press a lever and see on the screen a likely representation of the future event. But I could not now see the future event itself, since, as we have seen, of logical necessity I can only observe what is present or past. Such are the logical limits to time-travel.

When knowledge of a past state is not obtained by direct observation or by ready inference from its observable effects, but by laborious calculation from its effects or by inference from traces which are not effects, the knowledge is knowledge obtained by retrodiction. (If I infer from the track left on the barograph the day before yesterday that there must have been a storm yesterday, I am not inferring to the storm from an effect produced by it, but from an effect produced by a cause of the storm.) Knowledge of the future obtained from traces is knowledge obtained by prediction.

We noted at the beginning of the chapter that we depend for our knowledge of events and states at other places and temporal instants on reports of observers and on traces. Now there is a notorious asymmetry between the amount of our knowledge of the past and our knowledge of the future. The weather, the political situation, the number of road accidents in January of last year are things which I can easily find out. The weather, the political situation, the number of road accidents in January of next year are things I vainly try to guess. The asymmetry arises firstly because the reports which observers give of their observations are of what they are observing and have observed, never or hardly ever of what they will observe. (I say 'hardly ever' because some professed clairvoyants have given reports of what they are going to observe.) Assertions about what an individual has observed in the past can be checked by, and are normally corroborated by, the assertions of others. In general what I say happened is what you say happened. But others do not say the same about the future as I do and there are no checking procedures in use to see who is telling the truth about what will be. We can sum up this situation by saying that men have memory but not foreknowledge. Their non-inferential knowledge is of the past but not of the future.

In the second place the asymmetry is due to the fact that traces allow much easier inference to past events and states than to future events and states. Observation of a single state of one object of a size that men can handle often permits ready inference to past events. Footprints in the sand, tracks on tapes of tape-recorders, diaries, pottery, coins, etc., are clear evidence that certain sorts of events happened at some past and often precisely determinable instant. If there are human footprints in the sand now, then very probably within the last few days a man has walked there. If there is now a track of a certain kind on a tape of a tape-recorder, then very probably someone has sung a certain song within the last hundred years or so. But to spot traces of future events is very difficult. Even where we can do it, the process of inference to a future state is often a complicated one. To work out tomorrow's tides or what the vegetation will be in some region next year is a very complicated process, and to work out next year's political situation seems impossible. There are few easily recognisable traces of the future in nature which have any degree of reliability — the sun in the western sky signifying the coming of night is a rare example. Artificially some traces of the future have been produced — the track of the barograph and the tables in almanacs for next year's tides and eclipses are examples. But in general there are relatively few traces which allow easy inference to the future and those which exist are not very reliable.

For these reasons we know more about the past than the future. I next wish to examine whether and to what extent it is logically necessary that these factors should operate, that we should know more about the past than the future. Could there not be traces allowing easy inference to the future and claims about what will happen which were corroborated by others, the traces and the reports supporting one another? To examine this question, I shall consider the possibility of two different universes. In the first, corroborated reports are of the future only and traces allow easy inference only to the future. Knowledge of the past is as scanty as is knowledge of the future in our universe. In the second there are corroborated reports of the future and traces allow easy inference to the future, but corroborated reports of the past and traces allowing easy inference to the past remain as in our universe. The traces and corroborated reports of the future in both hypothetical universes are supposed to yield a similar amount of information to the amount yielded by them in our present universe about the past.

Let us begin by considering the first hypothetical universe. Could there be a universe in which traces allow easy inference only to the future, and reports about what will happen, and only such reports, were corroborated by others? Let us try to describe such a universe. Traces allow easy inference to the future. Footprints show where people will plant their feet; tracks on tape-recorders the songs that they will sing, and so on. Further, people make reports about what will be, and the reports which each makes agree in general with the reports which others make. There are, however, no traces allowing easy inference to the past, nor corroborated reports about what was. What people report without inferring and what people infer from traces back each other up, and establish the claims to knowledge by means of the other. Thus, men's claims about what they will observe are substantiated by the existence of present traces of the future events claimed to be going to occur. The evidence that some trace C_1 is of event D_1 about to happen, is not, as in our universe, that people agree that they have observed Cs being followed by D_s or other past phenomena, but that they agree that they will observe C_s being followed by D_s or other future phenomena from which could be inferred in a more complex way that D_s follow C_s.

Now it has been argued[1] that such a universe is not a logically possible one because it is logically necessary that we know more about the past than about the future. We mean by the past, the argument goes, that temporal period on one side of the present about which we are comparatively well informed and by the future that temporal period on the other side of the present about which we are comparatively ignorant. The degree of our knowledge about it is, on this account, the one and only way of distinguishing between past and future. (If we were equally well informed about the two temporal periods converging on the present, the concepts of 'past' and 'future' would presumably not be applicable.)

If this account is correct, then if we were no longer to distinguish between 'past' and 'future' by our degree of relative knowledge and so the description of the hypothetical universe no longer described the logically impossible, there would be no observable difference between the two universes. Our experiences in both should be the same. And if our experiences were the same, then, if we are now said to be better informed about one temporal period on one side of the present than about the other, it must be a matter of logical necessity that the former is the past, for no experience could show it to be so. Certainly we may in the hypothetical universe call that temporal period about which we are comparatively well informed 'the future' and the other one 'the past', but then we should just be using words in a different way. We would have changed the meaning of the words we use to describe the world, but the world would be the same.

The argument that if we no longer distinguish between 'past' and 'future' by degree of relative knowledge, the two universes would not differ in any observable respect runs as follows. In both universes the following would be the situation:

Temporal instants	t_1	t_2	t_3	t_4	t_5	t_6
Experiences	E_1	E_2	E_3	E_4	E_5	E_6
Corroboratable reports (supported by traces)	S_1	S_2	S_3	S_4	S_5	S_6

At t_1 we have experience E_1 and are able to give reports S_1 (which can be

[1] By Professor J. J. C. Smart [3] among others.

corroborated by others, and supported by traces) of our experiences at other instants, that is of E_2 to E_6. At t_2 we have experience E_2, and are able to give reports S_2 (which can likewise be corroborated by others and supported by traces) of our experiences E_3 to E_6 but not of our experience E_1. At t_3 we have E_3 and are able to give reports of S_3 of E_4 to E_6, but not of E_1 and E_2, and so on. Now in both our universe and the hypothetical universe, the situation is as follows. The reports we can give at any one instant about our experiences at other instants (reports which can be corroborated by others and supported by traces) can be arranged in an order according to the number of experiences of which we can give reports. In order of increasing knowledge the instants are t_6, t_5, t_4, t_3, t_2, t_1. The arrow marking the direction of decreasing knowledge runs from t_1 to t_6. Then in both universes it will be the case that at any instant, we can say (and reports of others and traces bear us out) that at any other instant we are in a position to report only on (and have knowledge by traces only of) events and states at instants more remote from the present in the direction of decreasing knowledge than that latter instant. Both universes will thus have the same formal pattern of knowledge of events at other instants. So, on this argument, if we do not distinguish between the periods 'past' and 'future' by the amount of knowledge we have then there will be no observable difference between the two universes. This argument does however assume, that the content of experience in the two universes does not differ in any marked way. However, a marked difference in content is to be expected from our attempted description of the difference between the two universes, and such a difference would give grounds for the inhabitants of our hypothetical universe to refer their knowledge of events at other instants to the future.

In the argument which I have expounded an experience is meant to be the ultimate atomic source of information from which we build up our picture of the world. The argument seems to suggest that experiences give us snapshots of the world at some infinitesimally small instant. But clearly if we are to have knowledge of the world, it is no use just having a collection of snapshots; we need to know the temporal order of their occurrence. So if experiences are the ultimate unit of information about the world, we must say that experiences are of things changing or remaining still, that is lasting for some significant temporal period. Thus, suppose I look into a water tank for a moment and see a ball moving. If my experiences consisted solely in a number of snapshots of the position of the ball, I would not subsequently be able to report whether it moved from left to right or right to left or jumped about discontinuously. So we must say that experiences must last for a significant period if they are to provide much information about the world.

This being so, there would be vitally important differences between the content of our experiences in the two universes. Thus, in our universe we often experience the unexpected. Normally in ordinary surroundings, and especially in strange surroundings, we see things, hear things, bump into things for which we were not ready. To us not expecting anything particular or perhaps something else, there comes information about what is in fact the case. In the hypothetical universe the information about the world would come to men ready for it. Experiencing the expected is a different sort of experience from experiencing the unexpected.

On the other hand, in the hypothetical universe there would be sudden forgettings, in the following sense. Once we had had an experience, we would subsequently be entirely unable to report on it. The loss of this ability would itself be a sudden experience, like the experience of being paralysed. In our universe such continual bathings in the waters of Lethe are fortunately unknown. So the content of experiences in the two universes would differ vastly, and this difference would give the inhabitants of the hypothetical universe grounds for saying that they had foreknowledge and not memory. What men reported at t_1 would include experiences at t_2, t_3 and so on looming up on them expecting just those experiences. They would also report a forgetting at t_6 of E_5, at t_5 of E_4, at t_2 of E_1. Men would thus know at t_1 that having an experience involved a sudden forgetting of it, and that it remained forgotten at all instants on the other side of it from t_1. If one loses the ability to report what is experienced as one experiences it, then the loss of knowledge must be immediately subsequent to the experiencing. Hence men in a universe in which there were no unexpected experiences, but, instead, sudden forgettings, would argue that they had foreknowledge and not memory; that is, would refer their knowledge of events at other instants to the future as we understand that. Hence the argument to show the logical impossibility of the hypothetical universe fails.

The argument was designed to show that if we do not distinguish 'past' and 'future' by degree of relative knowledge the hypothetical universe would have no observable difference from our own, and hence that our present relatively greater knowledge of the past cannot be a contingent matter, but a matter of logical necessity; and so that the hypothetical universe would not be logically possible. The argument failed because it failed to show that there would be no observable differences between the two universes, if we did not make this distinction.

However, the hypothetical universe is not a logically possible universe for a different reason: and, as a result of this, my attempt to describe it above will prove to have been incoherent. The reason is as follows.

The having of knowledge of what will be depends, as we have noted, on reports being made, corroborated, and supported by traces. Yet in such a universe the process of corroboration and obtaining support could never occur — for two reasons. The first reason is this. If we had foreknowledge, we would foreknow the conclusion of any process of establishing an item of knowledge. To establish whether an atom bomb is going to explode in Hull tomorrow, I consult with others and they report what will happen. I look at the alleged traces of the bomb and reach the conclusion that there will indeed be an explosion tomorrow. But (if we assume, as we have done, a similar amount of knowledge about the future to that in our universe about the past) I knew anyway that I was going to reach this conclusion, and what tests I was going to perform to establish it. Hence I establish it on the basis of a finite set of evidence known in advance. But the very process of finding out anything presupposes that any evidence may be relevant; and there is no limit to the argument that can be had about it; that new considerations can always turn up and be admitted as relevant or rebutted as irrelevant, that opponents can always suggest new tests of any theory, and that a theory will only be well confirmed if it predicts with success the result of tests unimagined when the theory was formulated.

None of this is present in the way things are found out in the hypothetical universe. We know the moves we are going to go through and the result which we will reach. But our situation then will be like that of a commission of inquiry knowing the conclusion it was going to announce and purporting to go through a certain number of investigations which it knew would support that future conclusion. Such a commission cannot be said to be finding anything out. Nor can we unless we are sensitive to unsuspected evidence. Yet, *ex hypothesi*, there cannot be such. Hence, we cannot be said to be acquiring knowledge. So our reports of what is to be cannot be said to be based on foreknowledge.

Secondly, not merely is the process of argument a fraud; but, when the conclusion is announced it cannot even be alleged to be announced on the basis of the evidence. For by the moment the conclusion is reached, the evidence will have been forgotten, and so will the point of the inquiry. So claims to knowledge cannot be tested in such a world. If claims to it cannot be tested, knowledge of events at other temporal instants cannot exist, for to describe what one has of them as knowledge presupposes that claims to it are testable.

For these two reasons I conclude that the description which I have attempted to give of such a world is incoherent. There could not be knowledge in such a world. Hence a world in which traces allowing easy inference and corroborated reports were of the future, and not the past, is not logically possible.

My next question is whether there could be a universe in which we had as much knowledge of the future as we now have of the past, added to our knowledge of the past. This would be a universe in which there were traces allowing easy inference and corroborated reports both of past and future. Neither sudden forgettings nor sudden gettings to know would occur.

Now is such a universe a logically possible one? While the second of the objections which I made to the possibility of knowledge in the previous hypothetical universe does not apply in any immediately obvious way, the first clearly does apply. In order to have knowledge, I must be prepared to meet new objections to and produce new evidence for my claims to knowledge. But I cannot do this in the universe being described, for I know in advance what arguments I shall produce, what objections will be made, and what conclusions will be reached about my claims to knowledge. My 'investigation' is that of a whitewashing commission and therefore improperly described as investigation and its results improperly described as knowledge. So a world in which we had 'knowledge' of the future similar in amount and additional to our 'knowledge' of the past would not be a world in which we had knowledge at all. The description defeats itself.

So far I have shown that a world in which we had knowledge of the future similar in amount to but instead of our knowledge of the past, and also a world in which the former knowledge was additional to the latter, are not logically possible worlds. This is because there must always be enough knowledge of the past for connected argument to take place and enough ignorance of the future for investigations to be carried out. But subject to this limitation, there seems no logical necessity preventing us from knowing very much more about the future and very much less about the past than we do. The question consequently arises whether there is any law of physics responsible for our comparative ignorance of the future or whether it is simply a matter of practical difficulty to find out much about the future. This question we shall investigate in Chapter 13.

So then we have seen in this chapter that there is a logical limit to the amount of knowledge we can have of the future and to the amount of ignorance we can have of the past if we are to have knowledge at all. There is also a logical limit to the kind of evidence we can have about the reliability of traces of the past. We cannot subject a claim that a trace is a trace of a past event to experimental test. There is a further logical limit to the way in which we can obtain knowledge of the future. Of logical necessity we cannot observe future events and states of objects.

There would seem, however, to be no similar logical limits to the amount and kind of knowledge we can have of events and states of objects due to mere spatial distance from ourselves. There is no logical necessity preventing me from using observers' reports and traces to learn about goings-on on any galaxy however remote. There may, however, be physical laws preventing this, and this issue we shall consider in Chapter 12. However, the more remote in space or time from ourselves are the states and events to which we infer from traces, the less reliable will our inference be.

BIBLIOGRAPHY

On the causal theory of time:
[1] H. Reichenbach, *The Philosophy of Space and Time* (trans. M. Reichenbach and J. Freund) New York, 1958, § 21.
[2] A. Grunbaum, *Philosophical Problems of Space and Time*, London, 1964, ch. 7.

On logical factors responsible for our greater knowledge of the past than of the future:
[3] J. J. C. Smart, 'The Temporal Asymmetry of the World', *Analysis*, 1954, 14, 79–83.
[4] J. E. McGechie, J. R. Searle, and Richard Taylor, Answers to Analysis Problem No. 9, 'Does it make sense to suppose that *all* events including personal experiences could occur in reverse?' *Analysis*, 1956, 16, 121–6.

On Time travel see, as well as the bibliography on the priority of cause to effect at the end of ch. 8, also:
[5] David Lewis, 'The Paradoxes of Time Travel', *American Philosophical Quarterly*, 1976, 13, 145–52.
[6] Larry Dwyer, 'Time Travel and Some Alleged Logical Asymmetries between Past and Future', *Canadian Journal of Philosophy*, 1978, 8, 15–38.

10 Times and the Topology of Time

In Chapter 2 we considered the question whether it was logically possible and whether we could have reason to believe that there was more than one space, and we concluded that the answer to both questions was positive. In this chapter we must ask the corresponding questions about time.

All instants temporally related to each other taken together constitute a time. An instant is temporally related to another if it is before or after the other. Could there (logically) be more than one time? As with spaces, Kant claimed that it was an *a priori* truth that there could only be one.[1] To say that there could (logically) be more than one time, is to say that not all periods of time, and so temporal instants, need be temporally related to each other. Another way of putting this claim is to say that not all events and states of objects (for brevity I shall normally refer only to events in future) need (logically) be temporally related to each other, viz. occur before, at the same time as, or after each other. Could there be events not temporally related to each other?

Now if it is a logically necessary truth that being temporally related is a transitive relation, then the only events of which we can have knowledge will be events which are temporally related to each other, and so we must believe that that there is only one time. This can be seen as follows. Any event about which we have knowledge, we now know about either because we are observing, or have observed, or will observe it or a trace of it, or because other observers report that they have observed it, are observing it, or will observe it, or a trace of it. Now if an observer observes an event, the occurrence of the event must, we saw in the last chapter, be simultaneous with or prior to the observation of it. If he knows about the event now because he has observed or will observe it, his present knowledge will be posterior or prior to the observation and the observation simultaneous with or posterior to the event. If the observer

---

[1] 'Different times are but parts of one and the same time.' [1], B.47.

knows about the event because he has observed a trace of it, the event must be prior to, simultaneous with, or posterior to the trace. For, as we saw in the last chapter, our usual grounds for taking an event A_1 as a trace of another event B_1, are that As always follow, are simultaneous with, or precede Bs. In these cases respectively A_1 will follow, be simultaneous with or precede B_1. If our grounds for taking A_1 as a trace of B_1 are less direct, they will be that it is the consequence of a well-established scientific theory that As are correlated with Bs and the same conclusion will follow. For a scientific theory, as we saw in the last chapter, states the consequences of some occurrence, what it brings about. Hence all the theory can tell us is that some event will follow another one, or two events will follow a common predecessor or something of that sort. So any event which is known, as a consequence of the adoption of a scientific theory, to be a trace of another event will be prior to, simultaneous with, or subsequent to it. Hence all events about which at a given instant an observer has knowledge occur at instants connected with the present instant by a temporal chain, viz. are before or after an instant which is before or after an instant . . . which is before or after the present instant. So if the relation of being temporally related is a transitive one, every event of which at instant t_1 a man has knowledge must be temporally related to, viz. prior to, simultaneous with, or posterior to possession of the knowledge at t_1. Further, if the relation of being temporally related is a transitive one, then since every event of which an observer has knowledge at t_1 is temporally related to his possession of the knowledge at t_1, every such event will be temporally related to each other. Hence the claim that there were events not temporally related to each other could have no evidence produced in its favour. For evidence would be evidence about events at other instants and the only ways in which we could learn about those events would be ways which presuppose that the events are temporally related to the event of learning about them.

If on the other hand, it is not a logically necessary truth that being temporally related is a transitive relation, then there could be three events or states E_1, E_2, and E_3 connected by temporal chains such that E_1 was after or before both E_2 and E_3 but E_2 and E_3 were not temporally related.

The question therefore arises whether it is a logically necessary truth that being temporally related is a transitive relation. We argued (p. 28), that being spatially related is of logical necessity a transitive relation. If a place A is spatially related to a place B and B to a place C, then A and C must be spatially related. For if A is spatially related to B, a traveller could, it is logically possible, travel by local motion from A to B, and if B is spatially related to C, a traveller could, it is logically possible, travel by local motion from B to C. Hence it is logically possible that a traveller could travel by local motion from A to C, if need be, by B. Hence A and C must be spatially related. A similar argument will not work, however, in the case of time. One does not travel by local motion between instants of time; and there is no obvious analogue in the case of time for this process.

Since, as we have seen, all instants about which at a given instant we have knowledge must be connected with it by temporal chains, evidence that there are two times could only have the form of evidence that a present instant will be followed or has been preceded by two instants which are neither prior to nor subsequent to one another. Could one ever have such evidence? To make plausible the suggestion that one might have such evidence, I shall recount a myth on the basis of which it might appear that men could claim with justification that a present instant had been preceded by instants not temporally related to each other and they by a common instant. If we accepted that the myth showed the latter, we would accept that it showed that there were two times; and hence we would have shown that it was logically possible that there could be two times. I shall, however, suggest that we ought not to accept this account of the myth, and that there is a conclusive objection to our doing so. I shall then urge that this objection must apply to any other two-time myth that could be constructed, and hence that we could never have knowledge of events belonging to two different times.

The myth is as follows. Two warring tribes, the Okku and Bokku, live in the land of Ug. The land of Ug is the only known inhabited land. The two tribes are assembled before their seer. He says to them: 'You have been fighting together too long and you clearly cannot live at peace with one another in this land of ours. I am reluctant to banish either of you from the land you love so dearly. However I have decided that while you may live in the same place as before, you must live completely separate existences. This will go on for the next twenty years as far as the Okku are concerned and for the next thirty years as far as the Bokku are concerned, the present members of the Bokku tribe being the more quarrelsome.' The seer then waved his wand. At this point, according to the Okku, the Bokku gradually disappeared from sight and the Okku inhabited the whole land and began to quarrel among themselves about who should own the territories suddenly vacated by the Bokku. The Bokku story is that the Okku gradually disappeared from sight and the Bokku inhabited the whole land and began to quarrel among themselves about who should own the territories vacated by the Okku. Later (twenty years later by Okku clocks and thirty years later by Bokku clocks), there happened a mysterious event differently described by the two tribes. According to the Bokku, the Okku suddenly reappeared in their midst; while the Okku claim that the Bokku suddenly reappeared in their midst. The members of the two tribes in the intervals of fighting over disputed territory then exchange their stories.

At the instant of reunion, according to both tribes, certain changes occurred in their buildings and possessions and to some extent natural environment. The changes, however, are differently described by the two tribes. The result of the changes is a compromise between the world known to the Bokku immediately before the reunion and that known to the Okku. During the thirty years since the separation of the tribes during which the Bokku lived, according to their account, in the land of Ug, they effected, on their account, certain changes to it. So, according to them, did the Okku, during the twenty years between the moments of separation and reunion when they inhabited the land of Ug. Both tribes claim to have built new houses and cut down old trees. Some of the houses claimed to have been built by each tribe remain after the reunion and some of the differences in the environment claimed to have been effected by each remain. Others do not.

Now notice the important features of this myth. Each tribe has a perfectly coherent, well-corroborated, and unbroken account of its history. There is no question of either account being the account of a mere dream. The Okku, for instance, have, they all say, always lived in the land of Ug. The landmarks are the ones that their fathers and grandfathers have told them about. Many years ago, the Okku say, the Bokku lived there too but they were so troublesome that a seer waved them away. Unfortunately he brought them back again only recently. The Bokku also claim a continuous history in the land of Ug, while claiming that it was the Okku who disappeared for a period. The detailed accounts given by the two tribes differ only in respect of the events of the intermediate period. Yet each account is well-authenticated, for all the members of each tribe agree with each other as to what happened.

Now what would be the evidence available to us as observers in the land of Ug after the reunion about where and when the events of the 'intermediate period' occurred? How are they to be fitted on to the spatio-temporal framework? The evidence of the Bokku and the Okku is that both series of events occurred in the land of Ug. If we accept this and also suppose that there was only one period of time, one series of temporal instants, between the instant of separation and the instant of reunion then we have to postulate the operation of some extraordinary physical laws. What happened at the instant of separation, a holder of this interpretation would urge, was that each tribe became invisible and intangible to the other and that implements wielded by one tribe had no affect on the other. The interpretation, however, would have to become very complex to deal with some phenomena. For instance, if an Okku chops down a tree at some instant t_1 on Okku clocks, no Okku can lean on it after t_1. However, a Bokku can lean on it until the reunion. By all Okku clocks the period of separation lasted twenty years; by all Bokku clocks thirty years. We could describe in detail how certain physical laws ensured that all the observers of one tribe suffered various visual and tactual illusions. But the story would clearly be tremendously complicated.

To avoid this awkward interpretation we might postulate that the two series of events occurred at different places from each other. This interpretation is not, however, a natural one. For the Okku and the Bokku are at all instants surrounded by qualitatively similar objects to those of the land of Ug before the instant of separation except for the members of the other tribe, and hence we have little grounds for saying that one or both tribes have changed their place. Normally, by the criterion of similarity (p. 21), if we find ourselves surrounded by qualitatively similar objects to those by which we were previously surrounded, except for a few which are no longer present, we judge (at least provisionally) that the few objects had moved, but that we ourselves had remained in the same place. In both stories indeed there are two mysterious events. At one instant, the other tribe disappears; at a second instant it reappears and there are a few odd physical changes to the environment at the same time. But the occurrence of two mysterious events in its history does not mean that a tribe has moved to a new place.

A further awkwardness for the two-place interpretation would arise if the Okku and Bokku both had evidence (in the way outlined in Chapter 2) that if the other tribe was situated at another place during the period of separation, that place would not be spatially related to their own place. We would then have to postulate that each lived in a separate space during the period of separation. Further, if we did really wish to claim that at the instant of separation one tribe moved to another space and returned to the land of Ug at the instant of reunion, would we not have to make the same claim about the other tribe? For the fact that both tribes experience equal upsets means that we have no justification for claiming that one tribe, say the Okku, had moved to another space, while the other tribe, the Bokku, remained put. In that case we would have to postulate the existence of three qualitatively virtually identical spaces. To claim that there must be two or three different spaces solely because a place ceased to be inhabited by a certain group of people seems to stretch the criteria of evidence for identity of place beyond those described in Chapter 1.

If we accept that to postulate that new and very odd laws of physics operate in the intermediate period or that the events experienced by the two tribes occur in two different places are both too complicated hypotheses, an alternative description in terms of two times is unavoidable. Time, on this interpretation, became split at the instant of separation into two streams Okku-time and Bokku-time, which were subsequently reunited. Each instant in Okku-time would be temporally related to each other such instant and to instants in Ug-time (viz. instants at which events occurred in the land of Ug before the separation and after the reunion) but not to any instant in Bokku-time; and conversely for Bokku-time.

If we refuse to allow the different physics or the different place interpretations, the two-time interpretation of the myth is inevitable. For take some event *O* in the story told by the Okku and another event *B* in the story told by the Bokku. Both events occur, we assume, in exactly the same place in the land of Ug. Now *O* cannot occur at the same instant as *B* — since whatever occurs at the same place and at the same instant as another event would (unless a radically different physics is postulated for the intermediate period), if it is the sort of thing such an observer observes (viz. the observer must have and use the senses and the categories appropriate to observing the event), be observed by any observer of the second event. Yet, *ex hypothesi*, *O* cannot be observed by any observer of *B*. Nor can *O* have occurred later than *B*. For in that case, if an observer of *B* stayed long enough at the place with his eyes skinned he would observe *O*. But, *ex hypothesi*, he would not. The same argument, *mutatis mutandis*, holds against the suggestion that *O* occurred earlier than *B*. So of two well-authenticated events *O* and *B*, *O* occurred neither before, nor at the same temporal instant as, nor after *B*. So, to generalise, *O* and all events in the Okku history of the intermediate period are not temporally related to *B* and all other events in the Bokku history of the intermediate period.

Now we have seen that in some respects the different physics and different place interpretations of the myth are very awkward, and the principle of simplicity might seem to lead us to choose the two-time interpretation. But I believe that to accept the two-time interpretation would be to deny a logically necessary truth and hence one of the other interpretations would have to be taken.

The logically necessary truth is that if it is logically possible for an unconscious material object to have a certain life history, viz. to be at various places for various different periods of time, then it is logically possible for a conscious material object to have the same life history. If a stove can be now here, now there; then it can be now here, now there, whether it has eyes or not. Its having eyes cannot make a certain life history logically impossible. Yet in order to maintain the two-time interpretation, we should be forced to deny this truth. For during the period of separation material objects must have a life history in both times. For unless the same material objects in the form of rocks and trees form the surroundings of the two tribes during the intermediate period, the two tribes would not be in the same place, and hence we would have to adopt the different-place interpretation of the myth. Conscious beings could not, however, exist in both times. For if a conscious being was simultaneously conscious of the events of both times, the two times would be the same time. If on the other hand he experienced only Okku events in Okku-time and Bokku-events in Bokku time, then any description of his experience at the instant of the reunion becomes self-contradictory. Since at both the instants of time, Bokku-time and Okku-time, immediately prior to the reunion, he did not know about the events of the other time, he must have come to know about the events of the other time at the moment of the reunion. But if he first came to know of the events of Bokku-time at the moment of reunion, of logical necessity he cannot have been experiencing the events of Bokku-time during Bokku-time, and similarly with Okku-time.

The same difficulty would arise with any other myth designed to show that we could have knowledge of events of two times. The myth would have to show, as we saw earlier, that men at some instant had knowledge of two events at past or future instants, which, they had reason to believe, were not temporally related to each other. Now did these events occur in the same space or not? If they did not occur in the same space, then however different they were from each other, men could easily attribute any peculiarities to the different laws of nature operative in the two spaces, as we have no reason to suppose that the same laws of physics operate in different spaces. Hence we need not suppose that they occurred in different times. If on the other hand the events occurred in the same space, then, relative to some frame F consisting of actually existing material objects they occurred either at the same place or at places at some distance from each other. The material objects forming the frame must exist in both times, for unless they did the events would not be occurring at any distance from each other. But then it is not logically possible that those objects be conscious. For, if a conscious being existed in both times, again any description of his experience at the instant of reunion becomes self-contradictory. I conclude that no myth designed to show that we could have reason to believe that there were two events temporally related to a third event but not to each other, and so that there were two times, will work; and hence that we cannot have any knowledge of any events in a time other than our own. We must, as Kant held, fit all events of which we have knowledge, into a single time stream. I have no argument to show that it is not logically possible that there be events which belong to different times; I argue only that no observer could have any knowledge of such events.

Time, as we know it, is one-dimensional. By this is meant that one temporal coordinate measuring temporal interval suffices to identify uniquely any instant in the one time of which alone we can have knowledge. To identify any instant you only have to say how long it was in the one time after or before some specified instant (such as the date conventionally ascribed to the birth of Christ or to the founding of the city of Rome). Could it be otherwise? To say that time was two-dimensional would be to say that temporal instants had to each other the relations of before and after in two ways. Of two instants t_1 and t_2, t_1 could be before t_2 in one direction yet after it in the other. For to say that time is two-dimensional is to say that some instants have this relation to each other. But this — of logical necessity — could not be. For if an instant t_1 is before another instant t_2 an event at t_1 could, it is logically possible (though not always physically possible because of such empirical factors as the finite velocity of the propagation of effects), produce an effect at t_2, but an event at t_2 of logical necessity could not produce an effect at t_1. Hence t_2 could not in any sense be before t_1. For if it were it would be logically possible for an event at t_2 to produce an effect at t_1. Time is, we must therefore conclude, of logical necessity one-dimensional.[1]

Time, like space, is of logical necessity, unbounded. After every period of time which has at some instant an end, there must be another period of time, and so after every instant another instant. For either there will be swans somewhere subsequent to a period T, or there will not. In either case there must be a period subsequent to T, during which there will or will not be swans. By an analogous argument any period which has a beginning must have been preceded by another period, and hence time is necessarily unbounded. Of course the Universe may not be temporally

---

[1] Kant claimed that this was a synthetic *a priori* truth — [1], B.47.

unbounded — after some instant t_1 there may be no Universe (this issue will be discussed in Chapter 15). But this is only true if there is some period after t_1 during which there is no Universe.

We saw in Chapter 8 that of logical necessity the same temporal instant never returns. Since time is of logical necessity unbounded, it must therefore of logical necessity be infinite. Since before every period of time having a beginning and after every period of time having an end there must be another period, and since the same instant and so period never returns, there is no limit to time. It has gone on and will go on forever. Space, as we saw in Chapter 6, is different. It is of logical necessity unbounded, but, as it may be the case that the series of places of some finite size is finite for in every series of such places we may always come to the same place again, space may or may not be infinite.

Space, we concluded in Chapter 3, would not exist if there were no physical objects in space. We have now reached the conclusion that time would exist without physical objects.[1] As Shoemaker [4] has argued, awareness of the passage of time involves awareness of change in or of objects. Minimally, this awareness may be simply awareness of a change in ourselves (e.g. me passing from a state of knowledge that it was a short time since so and so happened, to knowledge that it was a long time). But normally of course it will be awareness of things changing in the public world. Hence there could not be awareness of the passing of time without there being change in the world. But it does seem logically possible that there should be periods of time in which nothing changed and, Shoemaker argues, it would be logically possible to have evidence before or after its occurrence that there was a period of the world's history during which nothing changed. Shoemaker imagines a world divided into three regions, A, B, and C. In A something becomes motionless ('freezes') for a year every three years; this can be observed on most occasions from the other regions. In B everything freezes for a year every four years; and in C everything freezes for a year every five years; and normally these freezes can be observed from other regions. The observation of such regular patterns of change over many years would give observers good inductive evidence that these freezes would coincide in all three regions every sixty years; and so that there would be a year in which there was no change at all over the whole world.

In a similar way it seems logically possible that there should be a

---

[1] Our conclusion runs contrary to Augustine's claim that time began with the beginning of the Universe. He argued that 'Time could not exist, if there were no creatures', *Confessions*, 11. 30.

period of time in which there was nothing existent, preceded and followed by periods in which physical objects existed. One could have inductive evidence for the existence of such periods in a way analogous to the way in which Shoemaker suggested that one could have evidence for the occurrence of periods of time without change. (This has been suggested by Newton-Smith [6].) There could be a world, divided into three regions, *A*, *B*, *C*. On *A* physical objects vanish for a year every three years, after which objects similar to those which disappeared reappear. The objects in *B* vanish for a year every four years, and those in *C* for a year every five years, similar objects reappearing in the two regions after the year. These cycles of disappearance in one region can usually be observed from the other regions. The cycles of disappearance will coincide every sixty years. There would then be a period of a year in which there was nothing existent. Observers would have inductive evidence of the existence of such a period. The law which determined the appearance of new objects would state that objects similar to those which disappeared would appear in a place with the same spatial relations to qualitatively similar objects as those possessed by the old objects. (It could not state that they appear in the same place, relative to other objects, as the old objects which had disappeared; because every sixty years there would be no other objects, and so no such place.) Talk of periods of time during which or instants at which there are no physical objects makes sense because those periods and instants can be picked out by their temporal distance from periods at which there were changing physical objects. But in such periods in which there are no physical objects points of space cannot be picked out, because there are no frames of reference by reference to which points can be identified. It was for this reason that I argued in Chapter 3 that there could be no space without physical objects then existent; but I now argue that there could be time without such.

Time, then, being of logical necessity unique, one-dimensional, and infinite, has of logical necessity a unique topology. Instants have to each other the neighbourhood relations of points on a line of infinite length. Time having this topology cannot properly be said to have a metrical geometry as well. For a metrical geometry states, as we saw in Chapter 6, what relations of distance and direction and what propositions about area and volume are entailed by other relations of distance and direction. The temporal analogue to a point is an instant, to distance is temporal interval, to direction is being before or after. Time being one-dimensional, there is no temporal analogue to area and volume. Further, when we have identified temporal instants by their distance and direction from some specified instant, it becomes a matter of logical necessity what relations of distance and direction these have to each other. Thus for three instants a, b, and c, if b was an interval T_1 after a, and c was an interval T_2 after b, then c was after a an interval $T_1 + T_2$. That this conclusion follows of logical necessity from the premises is a consequence of the fact demonstrated in this chapter that if we know of two instants a and c both temporally related to an instant b, they must be temporally related to each other in a one-dimensional infinite time. Hence the only temporal path between two such points a and c must pass through b, and so the distance along it must be that from a to b plus that from b to c. But in contrast there are many paths between any two points of Space and hence it becomes an empirical matter where the one lies satisfying the criteria described in Chapter 4 for being a straight line.

We discussed in Chapter 9 methods for arranging events in temporal order, that is for ascertaining which occurs before another, which occurs at an instant prior to another. We do not, however, merely arrange events in temporal order, but measure the temporal interval between them. To do this we need a procedure for measuring temporal intervals, and such a procedure I shall call a time scale. Indeed it is often physically impossible to ascertain which of two events E_1 and E_2 was the earlier without the use of a time scale. Suppose for instance that E_1 and E_2 are spatially separate events, so far apart that it would be physically impossible for anyone who observed E_1 also to observe E_2 or to receive before observing E_1 signals from a man who had observed E_2, and conversely. In that case observers must judge which occurred first by judging which would have occurred first as indicated by readings on clocks at the place of each event, previously synchronised and ticking at the same rate. How time intervals are to be measured is the main theme of the next chapter.

BIBLIOGRAPHY

[1] I. Kant, *Critique of Pure Reason*, Transcendental Aesthetic, section 2, 'Time'.

On the possibility of two times see:
[2] Anthony Quinton, 'Spaces and Times', *Philosophy*, 1962, 37, 130–47.
[3] J. R. Lucas *A Treatise on Time and Space*, London, 1973, § 7.

On the relation of time to change and to the existence of physical objects, see:

[4] Sydney Shoemaker, 'Time Without Change', *Journal of Philosophy*, 1969, 66, 363–81.
[5] Bas C. van Fraassen, *An Introduction to the Philosophy of Space and Time*, New York, 1970, 11–30.
[6] W. Newton-Smith, *The Structure of Time*, London, 1980, Ch. 2.

11 Time Measurement and Absolute Time

What are the primary tests (in the sense described on p. 61) for temporal interval? To say that an event occurred n units of time after another event is to say that if a true clock recorded a certain reading at the place of the first event at the instant of its occurrence, then a true clock previously synchronised with the former would record a reading greater by n units at the place of the second event at the instant of its occurrence. Statements about temporal interval are statements about what would have been or would be the behaviour of true clocks.

A clock will be a true clock if it measures intervals by a true time scale. A clock measuring intervals in terms of one kind of unit, say hours, will measure time on the same time scale as another clock measuring intervals in terms of another kind of unit, say minutes, if all measurements of intervals between any two events by the two clocks are linearly related to each other, that is, each measurement by one clock in terms of its units is a constant multiple of the measurement of the same interval by the other clock in terms of its unit. A clock with only a minute-hand and a clock with only an hour-hand measure time on the same time scale.

But how are we to establish that a clock is a true clock? Any actual recurrent process — the rotation of the Earth, its revolution around the Sun — forms a clock, and yields a potential time scale whereby we can measure intervals. So too does any mechanical device which could be built — such as a pendulum clock — in which there is a recurrent process. Every occurrence of the process or some function thereof is assumed to occur after the same period of time, and we call this interval between any two occurrences — provisionally — a time unit. The rotation of the Earth takes a day. This gives us units in terms of which we can measure processes. If we assume that the Earth rotates at a regular rate against the background of the 'fixed' stars, we can divide the day into hours and minutes and seconds. On this basis using distance measurements as described in Chapters 4 and 5, we calculate the laws of nature. We find out how bodies move under the influence of forces of various kinds.

As with the parallel case, discussed in Chapter 4, for establishing that a body is a rigid body, there are two criteria for choosing the true time scale from among potential time scales. The first is that judgements of temporal interval between events which form part of our history and in particular of our personal experience must not diverge radically from those which we ordinarily make, by means of the day and year scales which have been in use for so many thousand years. Current events on Earth which last from when the Sun is at its highest point in the sky one day to when it is at its highest point the next day must be said to last approximately the same period of time as each other. So too must events which last from the period when the Sun is in Capricorn to the period when it is in Capricorn the next year. For if the adoption of a proposed time scale meant that last sidereal year was a thousand times as long as the previous sidereal year, that time scale could not be adopted — however simple the resulting laws of physics. This is so because such expressions as 'equal time interval', 'much longer time', etc., are given their meaning by the circumstances in which ordinary users of language judge it appropriate to use them. We only know what these expressions mean because we have been taught what these circumstances are. If physics radically contradicted all such obvious common-sense judgements about time intervals, we would have to conclude that it did not use 'time' in the same sense as did ordinary language. Certainly some law of physics may be simpler if the '*t*' in it is said to be a thousand times as great for last sidereal year as for the preceding one. But if we are to say that '*t*' denotes time, our judgements of *t*-interval must largely correspond to the judgements of time interval which would be made in ordinary circumstances of personal experience and, to a lesser extent, familiar human history by an ordinary user of language. Otherwise why say that the physicists' '*t*' denotes time?

The first criterion for a clock measuring the true time scale is a criterion which clocks need satisfy only approximately. Subject to it, we adopt as the true time scale at any place (by whatever frame identified) that scale by which an observer at that place will judge the laws of nature to have their simplest form. Let us make this point by showing how we find the true clock in the vicinity of the Earth by which to measure time interval. Our ground for not choosing the medieval town clock as the true timepiece is that relative to it the laws of nature would become highly complex. Days would be of uneven length, and so the Earth would be rotating at a significantly ever-varying rate. Bodies would attract each other with forces of intensity varying from instant to instant. The principle of scientific methodology that the scientist choose from among laws compatible with observed phenomena the simplest possible leads to the rejection of such a time scale. Since the simplest laws compatible with observed phenomena are those which the scientist has best reason to believe to be true, the time scale used in them will be that which the scientist has best reason to believe to be true. If we choose the scale given by the daily rotation of the Earth relative to the 'fixed stars' we find that — approximately — bodies obey Newton's three highly simple laws of motion and the neat inverse square laws of gravitation and electrostatic and magnetostatic attraction; or, possibly, other compara- tively simple laws, such as Einstein's field equations, to a yet higher degree of accuracy.

· Yet this time scale has to be corrected for it does not yield simple laws with perfect accuracy. Recently physicists have adopted as the official scale of time one based on the atomic clock. Atomic clocks have been developed since 1948, and have two essential parts. The first is a quartz crystal. Quartz has the property that if it is mechanically deformed a difference of electric potential is produced between its opposite surfaces; and conversely, the production of a difference of electric potential between its opposite surfaces brings about mechanical deformation. Because it has this property a quartz crystal can control the frequency of oscillation of an electronic oscillator. The crystal is adjusted at intervals by means of the second part of the clock, a caesium resonator. The quartz crystal is made to control the frequency of a radio wave. If this coincides with the natural frequency of a line of caesium, there occurs a transition in the magnetic properties of a beam of caesium atoms streaming through the resonator; otherwise the transition does not occur. Thus the frequency of oscillation of the quartz clock is checked by the natural frequency of radiation of an atom. If we assume that atoms invariably produce radiation of the same frequency, an unfailing check on the accuracy of clocks is available. One reason for supposing that this assumption holds near the surface of the Earth is that the laws of nature become simple and coherent, if judged by atomic clocks situated near the Earth. This is clear. For if judged by the atomic clock, the other laws of nature became much more complicated than when judged by some other standard, but we could easily account for the variation in frequency of atomic radiation with time as judged by the other standard, we would adopt the other standard. The other reason for adopting the assumption is that it confirms to some degree of approximation ordinary pre- scientific judgements about which processes last the same period of time. Scientific standards of time refine, but do not replace pre-scientific ones.

In 1966 scientists defined a second so that radiation of frequency

9,192,631,770 cycles per second was emitted by the $40 \leftrightharpoons 30$ transit of $\frac{133}{55}$ caesium. The role of this definition is, like the role of the similar definition for length, to correct and not to replace our normal methods of measuring time interval. For if by this standard tomorrow, though the day was in other ways similar to today, proved to be a thousandth of the length of today, we would say that the frequency of caesium radiation, and not the length of the day, had changed.

So then scientists adopt at each place that time scale which yields the laws of nature in their simplest form. At places in the vicinity of the Earth this means that they adopt an atomic clock. I shall later give reasons for supposing that clocks of the same construction should be used on all fundamental particles, that is, approximately, clusters of galaxies. I have so far assumed that study of physical phenomena would yield at any one place one true time scale. The possibility must, however, be admitted that some processes studied by science could be explained most simply and coherently using one provisional time scale; and other processes could be explained most simply and coherently using another provisional time scale. E. A. Milne once suggested that electromagnetic phenomena could best be dealt with by his t-scale, and dynamical processes by his τ-scale, and that these scales were related asymptotically.[1] Fortunately science has found no need to adopt his suggestion. Clocks on Earth can account for all different kinds of process successfully by the use of the atomic time scale and we have no reason to adopt a more complicated supposition for clocks in distant regions of the Universe. Hence, in the absence of evidence suggesting otherwise, we can talk of the true temporal interval at any place between two events and not merely of the temporal interval by this scale or that.

The account which I have given of the criteria for temporal interval is, like the account which I gave for distance, opposed to the conventionalist account given by Reichenbach and others.[2] In analysing the criteria for saying that two temporal intervals are equal, I have found that under normal circumstances these criteria yield clear un-arbitrary results. The

---

[1] E. A. Milne, *Kinematic Relativity* (Oxford, 1948), *passim* and especially pp. 224f.
[2] See Reichenbach [2] p. 117: 'All definitions . . . [of temporal interval] . . . are equally admissible'. Or Carnap: 'We cannot say that the pendulum is the 'right' choice as the basis for our time unit and my pulse beat the 'wrong' choice'. (Rudolf Carnap, *Philosophical Foundations of Physics*, London and New York, 1966, p. 83.)

same pattern of argument which I used against conventionalism about distance in Chapter 4 has force against conventionalism about temporal -interval.

So an observer at any one place can judge the time interval between events at that place. But what are the criteria of simultaneity of events at different places? How, in other words, is an observer to synchronise his clock with clocks at other places? Once we have a method for synchronising clocks at different places we can ascertain whether true clocks at two places yield the same time scale, viz. tick at the same rate. For we have only to measure temporal intervals on one clock and the intervals between the same two instants on the other clock (which we can do if we can ascertain when a reading on one clock occurs at the same instant as a reading on the other clock) and see if the intervals bear a constant linear relation to each other. If they do, the clocks at the two places yield the same time scale.

To synchronise clocks at the same place we have merely to adjust them so that they read the same. Synchronisation at the same place I shall term direct synchronisation. Synchronisation of clocks at two distant places I shall term indirect synchronisation. Intuitively, before we introduce any modern science, there would seem two possible methods of synchronising clocks at two distant places P and Q, the clock-transport method and the signal method. The clock-transport method is to have two clocks at P ticking at the same rate and then move one of them to Q and synchronise the clock at Q with it. This method is analogous to the method of measuring distances at places, where the distance at Q is given by the rod moved from P and held to be of the same length at Q as at P (see Chapter 4). The signal method is to send a signal at t_0 by P's clock from P to Q. It arrives at Q at t_1 on Q's clock. Immediately it arrives the signal is sent back from Q to P where it arrives at t_2 by P's clock. Now t_1 must lie between t_0 and t_2 by the principle that causes of logical necessity precede their effects. We then make the simplest hypothesis about the velocity of the signal coherent with the rest of physics. This hypothesis will yield a solution about where t_1 lies between t_0 and t_2, and hence provide a procedure for synchronising clocks at P and Q.

It would seem that one or other of these methods or perhaps both of these methods are the primary tests for synchronising clocks at different places, that is, they give the meaning of saying that a clock at P and a clock at Q have been adjusted to register the same time. A clock at P and a clock at Q will show the same time if a clock transported from P to Q, showing the same time as P's clock on departure, shows the same time as Q's clock on arrival and if a signal sent from P to Q and back again arrives at Q at the instant, on Q's clock, predicted by the simplest hypothesis about the journey and information about the instants at which, by P's clock, it left P and returned to P. I shall leave open for the moment the question whether both tests have to be satisfied for two clocks to be synchronised, or whether the satisfaction of one test suffices. We shall find that empirical considerations may help us to see what to say if one test is satisfied and the other not.

Unfortunately it is a physical fact that each method of indirect synchronisation leads to confusion in different circumstances. What I mean by this is that the results of different applications of each test conflict with each other or with the results of measuring time by true clocks at each place. Under such circumstances we cannot consistently maintain that the method in question is reliable.

Let us first consider the signal method. The signal method, it will be recalled, is to send a signal from some point P to another point Q and back again. The time of its arrival at Q, t_1 on Q's clock, must lie between t_0, the time of its departure from P, and t_2 the time of its return to P. If we had a signal which returned to P in no period of time at all, which was back again as soon as it went and so had infinite velocity, we could uniquely identify t_1. For then $t_1 = t_0 = t_2$. No such signal is known to science. In the absence of one, it is clearly desirable to choose the signal with the fastest two-way velocity, that is a signal whose total journey, there and back, for a given distance is the fastest. For such a signal will narrow down the possible values of t_1 as much as possible. The fastest signal known to physics is that of light and all other electromagnetic radiation, which has — at any rate within the region of our galactic cluster — a mean or two-way velocity of approximately 300,000 km/sec[1]. (We cannot of course measure the one-way velocity of light, i.e. its velocity relative to P on its journey from P to Q, without having first synchronised a clock at Q with one at P.) Now light (and all other electromagnetic radiation) has the peculiar property not shared by

---

[1] It has normally been held that the supposition that there could be particles or other signals which had a velocity greater than that of light would contradict the Special Theory of Relativity. However recently physicists have speculated whether after all there could be such particles to be called tachyons. But it looks rather as if the existence of tachyons would only be compatible with the universal applicability of Special Relativity if causes could follow their effects, and we have seen reason to suppose that that is not logically possible, or if logic is violated in some other way. See the discussion in John Earman, 'Implications of Causal Propagation outside the Null Cone', *Australian Journal of Philosophy*, 1972, 50, 222–37, and the articles referred to therein.

material objects that — at any rate in the region of our galactic cluster — its mean velocity has the same value in all inertial frames. By this is meant that when a light signal is sent between any two points P and Q and back again (stationary or in motion relative to each other), and that passage is marked as a passage on inertial frames moving relative to P or Q with any velocity, then the mean velocity of the signal will be the same in all such frames.

That the mean velocity of light has the same value in all inertial frames is the most natural extrapolation from a very large number of experimental results, the most celebrated of which is the Michelson— Morley experiment. It was assumed by Einstein in formulating the Special Theory of Relativity and confirmed experimentally in 1913 that the velocity of light received at E from a distant source S is independent of the velocity of S relative to E. Light from each of a pair of double stars, one of which is receding from the Earth and one of which is approaching it (judged by the criteria of Chapter 5) has the same one-way velocity over the surface of the Earth. This can be shown by reflecting flashes of light from the two stars which arrived simultaneously at P on Earth to another point Q on Earth. The two light flashes will arrive at Q at the same instant. So, the Earth being to all appearances a typical near-inertial frame, either the mean velocity of light is the same in all inertial frames or it varies with the velocity of the receiving frame E relative to something else other than the source. During the nineteenth century it was supposed by many that light had a constant one-way velocity relative to an underlying medium, ether, and that its velocity relative to E varied with the velocity of E relative to the ether. The Michelson—Morley experiment was undertaken in 1887 to ascertain the velocity of the Earth relative to the ether. A light signal was sent from a source (O) stationary on Earth a distance L (relative to the Earth) in perpendicular directions to mirrors M_1 and M_2 stationary on Earth and reflected back again to the source (O). Suppose that along OM_1 the Earth has a velocity V relative to the ether. Then if light has a constant one-way velocity c relative to the ether, and it takes an interval T_2 to travel in a direction perpendicular to the direction of the Earth's velocity relative to the ether and back again, $T_2 = \dfrac{2L}{\sqrt{c^2 - V^2}}$. In the direction the same as that of the Earth's velocity relative to the ether and back

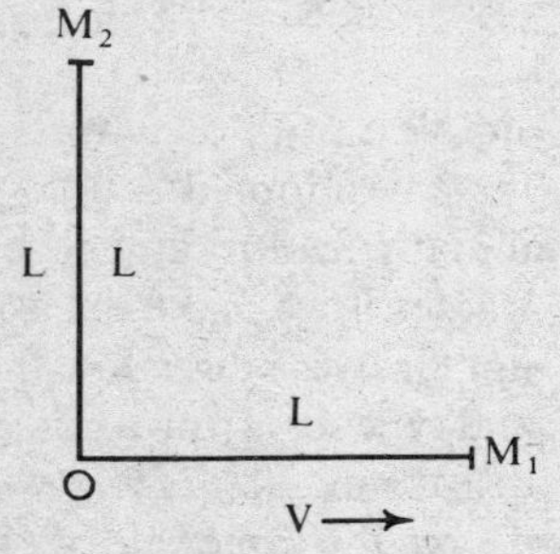

Figure 8

again, it will take an interval $T_1 = \dfrac{2Lc}{(C^2 - V^2)}$. The difference between

the two intervals should be detected at O from the interference fringes
produced at O by combining the two beams. However, in whichever
direction OM_1 was aligned no interference fringes resulted. Hence there
was no direction in which the Earth had a velocity V relative to the ether.
So either the Earth was stationary relative to the ether at the moment of
the experiment or the mean velocity of light was always constant on the
Earth, and independent of the Earth's velocity relative to any underlying
medium. The same negative result of the experiment was obtained at
many different periods of the year. Hence either the Earth was
permanently fixed in the ether while Sun and planets revolved around
it — a supposition ruled out since Galileo (see Chapter 3) — or the mean
velocity of light was constant relative to the Earth with whatever velocity
relative to any other body or medium known to science in whatever
direction it moved.

The Earth being to all appearances a typical near-inertial frame,
proponents of the Special Theory of Relativity have drawn from the
experiment the conclusion that the mean velocity of light is constant in
all inertial frames. This conclusion is in harmony with so much scientific
theory (especially the previously formulated electromagnetic theory) and
so many other subsequent experiments that the evidence for it may be
said to be very strong — at any rate for inertial frames within the region
of our galactic cluster (the question of the velocity of light on the cosmic
scale will be considered later). The most natural simplification of the
conclusion — although one which we shall later give reason for
abandoning — is that the signal has in every inertial frame its mean
velocity as its one-way velocity relative to the source in every direction,

viz. $t_1 = \dfrac{t_0 + t_2}{2}$. Since, as we have seen, science aims to formulate the

simplest laws of nature compatible with phenomena, the assumption
must be adopted — unless it leads to complexities elsewhere. It does not
initially lead to any such complexities. On the contrary, on all inertial
frames if we make the assumption about the equality of the velocity of
light in every direction, then all other signals (e.g. a bullet of a certain size
and type sent from a gun of a certain construction) have, in the absence of
disturbing factors (e.g. wind) detectable by other effects, relative to their
source the same velocity in every direction. Hence, if we judge time at Q

by the formula $t_1 = \dfrac{t_0 + t_2}{2}$, using other signals than light, we would in

the absence of disturbing factors get the same result.

In order to synchronise his clock with the clock at P by the signal method, an observer at Q must know when a signal arrives from P the instant of its despatch and the instant at which the signal which he then despatches will arrive back at P. Only if he knows both t_0 and t_2 when his clock reads t_1 will he know what the clock reads at P at that instant. Information about the instant of despatch may be known in advance or could be transmitted with the signal. There are various easy methods of ascertaining in advance when a given signal will arrive back at P, as judged by the clock there. Thus we can send a series of signals at equal intervals from P as judged by P's clock. We can note the formula governing their arrival back at P and so predict of the next signal when it will arrive back at P. If Q is stationary or in uniform rectilinear motion relative to P, the signal will arrive back at P at equal intervals. By this method an observer at Q can continually check his clock against the clock at P and so it will appear whether or not the two clocks tick at the same rate, viz. yield the same time scale.

Now the signal method of synchronisation leads to no confusion for clocks stationary relative to each other, that is, situated on the same frame of reference — so long as we are considering only one set of such clocks. For any number of such clocks stationary *inter se* the signal method enables clocks to be synchronised by a method which gives clear and unambiguous results. Thus if we synchronise a clock at P with one at Q and one at Q with one at R, we shall find that the one at R is synchronised with the one at P. Using this standard we can check whether true clocks at each place on the frame yield the same time scale, that is whether natural processes run at the same rate at each place on the frame. We find that on inertial frames true clocks, viz. atomic clocks, yield the same time scale everywhere. On other frames this is not necessarily so. Thus on a frame rotating relative to an inertial frame, clocks on the circumference go more slowly than clocks at the centre.

But the moment we consider two sets of clocks in relative motion, the members of each set being stationary relative to each other, the signal method of synchronising clocks stationary relative to each other leads to confusion.

Given the signal method of synchronisation for clocks on each frame, and given that light has the same one-way rectilinear velocity in all inertial frames, we can deduce the celebrated Lorenz transformations which form the core of the Special Theory of Relativity from which its other results are derived. If an inertial frame F' is moving with uniform velocity v along the x-axis relative to inertial frame F, then an event at (x, y, z, t) in the frame F will have coordinates (x', y', z', t') in F' (the first three coordinates being Cartesian spatial coordinates and the last coordinate the temporal coordinate, spatial and temporal origin points being the same for measurements in both frames) such that:

$$x' = \beta(x - vt)$$
$$y' = y \qquad t' = \beta\left(t - \frac{vx}{c^2}\right), \text{ where } \beta = \frac{1}{\sqrt{1 - \frac{v^2}{c^2}}}.$$
$$z' = z$$

Given the signal method of synchronisation, the Special Theory of Relativity which purports to hold in a pure form only for a universe empty of matter, is found to hold in all respects to a very high degree of approximation within the region of our galactic cluster except near regions of very high matter density.

Now suppose two momentary events E_1 and E_2 say flashes of lightning, to occur. E_1 occurs at point P on F and point P' on F'. E_2 occurs at point Q on F, and point Q' on F'. Now it is a consequence of the Lorenz transformations that if E_1 and E_2 are simultaneous as judged by the light signal method used in F, they will not be simultaneous as judged by the light signal method used in F', and conversely. For if the time interval between E_1 and E_2, as judged in F, is $t = 0$, then the time interval between them, as judged in F' will be $t' = -\dfrac{\beta vx}{c^2}$. Conversely if t', the time interval between E_1 and E_2 as judged in F', $= 0$, $t = \dfrac{vx'}{\beta c^2}$.

Consequently the signal method for synchronising clocks leads to confusion. For if it is valid, you find of two events that they both are and are not simultaneous. The relativity of simultaneity, as judged by the signal method is, as we have seen, predicted by the Special Theory but the formula $t' = \beta\left(t - \dfrac{vx}{c^2}\right)$ has been well confirmed independently.[1]

There are exactly two possible ways out of this awkward situation. The way adopted by Einstein and expounded in text-books on the Special Theory of Relativity is to say that the ordinary concept of simultaneity could only be applied if two events simultaneous by the light signal method in one frame were simultaneous in all equibasic (viz. inertial) frames, and so we must substitute for it the concept of simultaneity in a frame.[2] Two events are simultaneous in a certain frame if the clocks in

[1] By the Ives–Stilwell effect and cosmic ray phenomena. For details see [3] pp. 213–15.
[2] See A. Einstein, 'On the Electrodynamics of Moving Bodies' (originally published 1905) in A. Einstein *et al.*, *The Principle of Relativity* (W. Perrett and

that frame synchronised by the light signal method show the same reading when the two events occur. So E_1 and E_2 may be simultaneous in F, but not in F'. But no contradition arises, since it does not follow from E_1 and E_2 being simultaneous in F that they will be simultaneous in F' or any other frame.

The alternative is to deny that, except possibly in one preferred frame, the light signal method measures simultaneity. We could hold for instance that simultaneity is properly judged by the light signal method in a certain preferred frame F only, and that two events simultaneous in F are truly simultaneous. Then to find out what the time interval is between any two events we have to find out what judgement about time interval would be made by an observer on F using the light signal method. This means assuming that only in F is the velocity of a reflected light signal constant in both directions. In all other inertial frames it will have some velocity j (different for each frame) in a certain direction and $\dfrac{cj}{2j - c}$ in the opposite direction, with intermediate velocities in intermediate directions. This will mean that its mean velocity for a two-way trip will be the same in all inertial frames, as observable. This interpretation of physics can be carried through perfectly consistently,[1] but the trouble is, if we confine ourselves to the Universe of Special Relativity, viz. a Universe virtually empty of matter, that there are no grounds for choosing one inertial frame rather than any other as the preferred frame. The reinterpretation of physics leading to a unique measure of simultaneity can be carried through perfectly consistently and with identical complexity for any choice of preferred inertial frame, and given the applicability of the Special Theory, any non-inertial frame would be a less basic frame and so one which science ought not to use. So unless we have grounds for adopting a scientific theory other than Special Relativity, and on the basis of it for judging that for the laws of physics, or perhaps just for the laws of physics in some region, there is a most basic frame of reference, we have no justification for adopting any particular version of this alternative. If, however, we were to find a most

---

G. B. Jeffery) (London, 1923) pp. 42f. 'We see that we cannot attach any *absolute* signification to the concept of simultaneity, but that two events which, viewed from a system of coordinates are simultaneous, can no longer be looked upon as simultaneous events when envisaged from a system which is in motion relatively to that system.'

[1] For the first stages of such a reinterpretation and demonstration of its consistency see Grunbaum [4] ch. 12, section B.

basic frame of reference, we ought to take this as a preferred frame (since the laws of physics ought to be referred to most basic frames) and so we would have a coherent standard of simultaneity, a method of synchronisation which does not lead to confusion.

So much for the signal method, the circumstances in which it does and the circumstances in which it does not lead to confusion. We must now consider the clock-transport method which seems perhaps more obviously a primary test for indirect synchronisation than the signal method. This method, it will be recalled, for synchronising clocks at *P* and *Q* is to have two true clocks synchronised at *P*, and then to move one to *Q*. The clock at *Q* is then synchronised with the clock moved from *P*, and the clocks at *P* and *Q* are then said to show the same time. If we moved clocks from *P* to *Q* at regular intervals we could then see whether clocks remaining at *P* and *Q* yielded the same time scale.

Now the clock-transport method normally leads to confusion, and this for two reasons. The first is that clocks showing at *P* the same reading taken from *P* to *Q* by different routes or at different velocities will record on arrival at *Q* different readings. The second difficulty is that a clock moved from *P* to *Q* and then taken back again will not in general record the same passage of time as a clock which remained at *P*, but a shorter time. (Here the celebrated twin 'paradox' that if one twin remains on the Earth while the other travels in a space rocket on a long journey and then returns to Earth, the latter will have aged less than the former. For his ageing mechanism would be expected to operate at the rate of a clock transported with him.) Hence either when moved from *P* to *Q* or when moved from *Q* to *P* it must have shown on arrival a smaller reading than that shown at the same instant on the clock at the point which it left. So we can only use the clock-transport method if there is a route and velocity of moving a clock which is such that moving a clock from *P* to *Q* and back again has the consequence that on arrival back at *P* it records the same time as the clock which stays at *P*, and which is such that we have reason to suppose that it ticks at the same rate on its journey to *Q* as on its journey back. Fortunately, given the Special Theory of Relativity, within any one inertial frame there is such a method of moving clocks so that this holds when the smallest of corrections is made to the clock. The method is to move the clock (by any route) very slowly, relative to that inertial frame. Noting that when we move the clock from *P* to *Q* and back very slowly, its reading on arrival back at *P* diverges very little from the reading on the clock which stayed at *P*, and the more slowly it is moved, the less it diverges, we can work out the limit of this process — i.e. to talk slightly metaphorically, what it would read if it were moved infinitely slowly. That reading will exactly coincide with the reading on the clock which stayed at *P*. The simplest hypothesis is to suppose that the clock ticks at the same rate when moved from *P* to *Q* as when moved in the other direction. Hence we have reason to use the clock-transport method of infinitely slow transport in a straight line in order to synchronise a clock at *Q* with one at *P*. Ellis and Bowman [9] point out that, given the Special Theory of Relativity, this method gives exactly the same results as the signal method so far described; and in opposition to the 'conventionalist' approach of [10], Ellis [11] rightly urges that all this is good reason for adopting the judgements of simultaneity which these methods yield. However, as we have seen, all this holds only within a given inertial frame; and if we are to use a standard of simultaneity, we need grounds for preferring one inertial frame to others.

Leaving for the moment these difficulties about synchronisation, let us next present and give the evidence for the overall picture of the Universe, given by modern cosmology. In so doing we shall at last present the evidence for the cosmologist's thesis about basic frames which I set forward in Chapter 3.

An observer on Earth notes that, on the evidence of their Doppler shift, all other clusters are in recession from his own. All other clusters at any given distance (estimated by the methods described in Chapter 5) in whatever direction have, when allowance is made for small random velocities, the same velocity of recession from our cluster, a velocity which increases with distance (possibly uniformly), viz. the observed region of the Universe is, relative to our cluster, approximately isotropic. All observed clusters belong to one of a few types and the different types are spread throughout the observable region. Now is it only relative to our cluster at this instant on its clocks that this isotropic recession of observed clusters occurs, or does it occur relative to the other clusters at all instants on the clocks of each? The supposition that only relative to our cluster at this instant is there isotropic recession seems unreasonable, and the principle of simplicity dictates us to postulate that for all clusters at all instants on the clocks of each the observable part of the Universe is approximately isotropic (viz. observed clusters at any given distance from any observed cluster recede from it with approximately the same velocity). We can avoid the awkwardness that the isotropy is only approximate by postulating an imaginary frame of reference in the vicinity of each cluster, relative to which the recession of other such frames is uniform. Such a frame is termed in the literature of cosmology a fundamental particle, and an observer imagined as situated on one a fundamental observer[1].

The fundamental particles are then in uniform recession relative to each other. A cluster may rotate relative to its fundamental particle, but any velocity of recession of the cluster from the particle will be small and temporary. Motion of a cluster relative to its fundamental particle is random and thus equally likely to be in any direction. The supposition that we can postulate associated with each cluster a fundamental particle its motion relative to which satisfies these conditions, and that relative to any fundamental particle at any instant the observable Universe is isotropic, may be called the principle of isotropy. That it holds is an empirical postulate which might turn out to be false. Wireless messages might one day be received from astronomers in a distant cluster reporting that relative to them the observable Universe is very far from isotropic. But until such messages are received it seems reasonable to adopt the principle.

If then the observable Universe is always isotropic relative to each fundamental observer, it seems a further simplification to postulate that the laws of physics as formulated by an observer on each fundamental particle, using his measures of distance and clocks, are the same; and so that they are the same on each cluster, if we make allowances for any random motion or rotation of the cluster. This principle we shall call the principle of equivalent laws. By it the law that a certain machine will tick always at the same rate will hold on every cluster and so true clocks on every cluster will have the same construction. The principle is an empirical postulate which might turn out to be false. There is, however, some empirical evidence that the principle is true, at any rate for local laws, that is laws governing the behaviour of matter on each cluster. One piece of evidence is that light spectra of distant clusters can most reasonably be interpreted, as we saw on p. 91, as spectra of elements

[1] Some recent research suggests that the isotropy *may* indeed be only approximate. Recent measurements indicate that relative to the background cosmic radiation (On this see Chapter 13), our galactic cluster is moving in space in the direction from Aquarius to Leo with a velocity of perhaps as much as 600 km/sec. For an elementary account of this work and references to more technical literature, see Richard A. Muller "The Cosmic Background Radiation and the New Aether Drift', *Scientific American*, May 1978, 64–74. But even if these results are accepted, they still leave open the question whether it is our galactic cluster which is moving away from our fundamental particle or the background radiation which is moving away from it.

known to us to be shifted to the red. If the interpretation is correct then elements on distant clusters have the same frequency relative to each other as elements known to us, and hence an atomic clock graduated by the natural frequency of a line of one element there would ensure such simple laws of nature as the constancy over time of the frequency of light emitted by the other elements.

Once we have adopted, as we ought to adopt on grounds of simplicity, the principles of isotropy and equivalent laws, we conclude that relative to each cluster which we observe there are clusters observed receding isotropically, on each of which the same laws of physics hold. We can thus use observed clusters for which the principles are satisfied as traces that beyond them there are other clusters for which the principles are satisfied, and so we can extrapolate spatially *ad infinitum*.

It follows from the principle of equivalent laws that the fundamental particle in the vicinity of each cluster is for the laws of cosmology an equibasic frame, viz. proposed laws of cosmology acquire their simplest form relative to any fundamental particle, but no fundamental particle is more basic than any other one.

However, cosmologists suppose that for setting forward the local laws of physics, the local fundamental particle forms the most basic frame of reference. Their grounds for supposing this are that relative to any other fundamental particle the local laws would prove more complicated. For instance they have reason to believe, as will be shown on pp. 199f, that the mean or two-way velocity of light, locally constant, is variable relative to distant fundamental particles. If the cosmologists are right, then there is in the vicinity of each cluster a most basic frame of reference relative to which the laws of physics ought to be formulated and hence judgements of distance and temporal interval to be made. The difficulty with the signal and clock-transport methods was, we saw (pp. 188f), that before we could make such judgements we had to specify a preferred frame of reference. If Special Relativity, which is a theory of an empty Universe, were our theory of mechanics we could not do this. But the distribution of matter suggests a preferred frame in the region of each cluster, and hence these methods can be used for synchronising clocks within the cluster, on the supposition that only relative to the fundamental particle does light have its mean velocity in both directions. The two-way velocity of light remains approximately constant relative to all approximately inertial frames in the vicinity of the cluster, but the one-way velocity varies with the direction in which the frame is moving relative to the fundamental particle. The use of these methods, incidentally, would be expected to reveal that clocks on the edge of galaxies rotating relative to their fundamental particle go more slowly than clocks of identical construction in the centre. This is a prediction of Special Relativity which still applies to a high degree of approximation in such small regions of the Universe as galactic clusters.

Clusters themselves, however, are in mutual recession. Consequently the light signal method could afford no unique judgement of simultaneity for events on different clusters. An event E_1 on one cluster C_1 and another event E_2 on another cluster C_2, which, by the light signal method using the fundamental particle associated with C_1 as its frame of reference, were simultaneous, would not be simultaneous by the light signal method using the fundamental particle associated with C_2 as its frame of reference. And there would seem no preferred method at present of moving clocks between galaxies. However, the constitution of the Universe described above does suggest a preferred method of clock-transport which will lead to no confusion — on the following grounds. By the principle of equivalent laws the laws of physics formulated by an observer situated on each equibasic frame, that is in modern cosmology each fundamental particle, will be the same. Hence *either* each process of a kind on each equibasic frame, including the ticking of clocks, occurs at a rate generally different from a process of the same kind on another equibasic frame, while on each each process of one kind has the same rate relative to each process of another kind, *or* each process of the same kind occurs at the same rate on each equibasic frame. Clearly the simplest supposition to make and hence the one which we ought to adopt in the absence of counter-reasons is the latter. This principle we shall call the principle of similar clocks. If there are no objections to adopting the principle, then there will be a unique method of ascertaining the instant by the clocks on any cluster at which a distant event occurred. For clusters of galaxies are now receding from our cluster. So by the principle of isotropy either the mutual recession of clusters has been going on since an instant by the clocks of our cluster when all clusters were very close together or it has been going on since an instant by the clocks of our cluster when all clusters were approximately stationary relative to each other.[1] But in either of these cases clocks on each cluster could have been synchronised at the instants referred to. In the first case clocks would

---

[1] If one of these two properties holds of the Universe, Weyl's postulate will be satisfied (see H. Weyl, 'Redshift and Relativistic Cosmology', *Philosophical Magazine*, 1930, 9, 936–43). Weyl's postulate is that the clusters lie in Space-time on a bundle of geodesics diverging from a point in the finite or infinitely distant past.

have been very close together and so could have synchronised directly by comparison at what would be, approximately, the same place. In the second case when all clusters were approximately stationary relative to each other, all equibasic frames would have formed the same one most basic frame and hence there would have been a preferred frame and so synchronisation could have taken place by either method relative to it, viz. relative to every fundamental particle.

So if the principle of similar clocks can be maintained, we have a method for ascertaining the instant on our clocks at which a distant event occurred. It occurred at the instant shown by clocks of the fundamental particle in its vicinity, and the clocks of that particle have ticked at the same rate as the clocks of our fundamental particle since the instant at which they could have been synchronised with them. Hence if by the clocks of its fundamental particle the event occurred 5,000 million years since its cluster was close to our cluster, then it occurred 5,000 million years after that event by the clocks of our cluster.

Can the principle of similar clocks be maintained? In the Universe of Special Relativity it could not be maintained. Here the equibasic frames are the inertial frames. Now consider three inertial frames A, B, and C all moving with uniform rectilinear velocity relative to each other. At the initial instant of the experiment B passes A, and observers on each synchronise clocks directly; later B passes C moving towards A, and observers on B and C synchronise clocks directly; finally C passes A, and observers on both note the readings on the two clocks. It is a consequence of the Special Theory that the clock on C will record a smaller passage of time than that on A. Consequently the principle of similar clocks is false. According to Special Relativity if two clocks are moved symmetrically relative to an inertial frame they will record on reunion the same passage of time (an example of two clocks moved symmetrically relative to an inertial frame would be two clocks moved each in a different circle of the same radius with the same velocity meeting again at the same point on an inertial frame). But it is also a consequence of Special Relativity that a clock which returns to the same point on an inertial frame is not itself an inertial frame.

These difficulties do not arise in the Universe of modern cosmology. For its equibasic frames are not inertial frames, but fundamental particles, and the relative motion of these is very different from the relative motion of equibasic frames on Special Theory. For the only instants, according to modern cosmology, at which a fundamental particle meets another fundamental particle are, because of the principle of isotropy, the instants, if any, at which it meets all fundamental

particles. Hence an experiment of the type described above could not be
performed if A, B, and C were fundamental particles. Further, the motion
of any two fundamental particles A and B at some distance from each
other is always symmetrical relative to a third fundamental particle, viz. a
particle lying at each instant on its clock midway between them. If
Special Relativity were valid in this field, it would be a consequence of it
that since C is approximately an inertial frame, the clocks on A and B tick
at approximately the same rate. Special Relativity is not, however, valid
in this field. However, since the motion of all fundamental particles is
symmetrical relative to another one, relative to which by the principle of
isotropy the Universe is symmetrical, there would seem no grounds for
supposing that one clock goes more slowly than another. Considerations
of simplicity therefore lead cosmologists to adopt the principle of similar
clocks.

This principle, like the two principles previously discussed, is empirical. This can be seen by the fact that there are empirical tests which
could — it is physically possible, though hardly practically possible
today — conclusively show the principle to be wrong, if in fact it was
wrong; although it would be less easy to falsify conclusively than those
principles would be. First the principle could conclusively be shown
wrong if clocks previously synchronised directly when they were at
approximately the same place were to meet again. For then it could be
seen whether the same interval had passed on each. There are a few
cosmological theories of oscillating Universes which allow such tests. A
second possibility of conclusive falsification also arises if the clocks had
originally been synchronised directly. Suppose a signal sent from one
cluster C_a at t_0 to another C_b. Let it arrive at C_b at t_1 and at that instant be
reflected back again so as to arrive at C_a at t_2. If t_1, as measured by the
clocks of the fundamental particle associated with C_b, did not lie between
t_0 and t_2, as measured by the clocks of the fundamental particle
associated with C_a, then the clocks cannot have been ticking at the same
rate. If the original synchronisation was by the signal method when the
clusters were stationary relative to each other it would be less easy to
prove the principle of similar clocks false if in fact it was false. One way
would arise by sending a subsequent signal as above. But if the instant of
its arrival at C_b did not lie between t_0 and t_2, it might be possible to avoid
the conclusion that the principle of similar clocks was false by supposing
that the original synchronising signal did not have its mean velocity in
both directions and hence the clocks had not been originally synchronised correctly.

It can be seen that such conclusive falsification of the principle of similar clocks, if it were false, though physically possible, would not be very easy to achieve. Hence we must rely on more indirect counter-evidence if we are to show it false. However in the absence of such counter-evidence it ought to be adopted on the basis of the slender evidence earlier adduced. If we adopt it, we have now a unique method of clock-transport which is to be preferred to other methods and which will, as far as we can tell, lead to no confusion. The method is to keep clocks on their clusters; to move clocks, in the cosmologist's phrase, with the mean motion of the matter in their vicinity. The method can only be used for synchronising clocks not on the same cluster, for within the region of each cluster stars or groups of stars do not form equibasic frames moving isotropically relative to each other, and so there is no justification for adopting the principle of similar clocks for that scale. Previously we had supposed that we could find out by a method of synchronisation whether or not clocks at different places ticked at the same rate. Rather it is the case on the cosmic scale that we adopt, with some justification, the supposition that they do, and use this to justify a method of synchronisation.

The time scale which would be measured by true clocks on each cluster originally synchronised at some past instant is known as the cosmic time scale. We can refer to any event in the history of the Universe by giving the cosmic instant at which it occurs. This means that we can say of any two events E_1 and E_2 on different clusters C_a and C_b whether or not they are occurring at the same instant as, before or after each other. We calculate the time interval by the clocks of each cluster elapsed since an instant at which clocks on C_a and C_b could have been synchronised. Whichever event occurred after a longer time interval since that instant is the later.

It is a consequence of the principles of isotropy and equivalent laws described earlier and the existence of a cosmic time scale that the density, energy and distribution patterns of physical objects (viz. the clustering of galaxies and their velocities of mutual recession) are approximately the same in all regions of space of more than a certain volume of size at any given cosmic instant[1] (these regions will be regions of such a size that the

---

[1] The claim that the density, energy, and distribution patterns of physical objects (viz. the clustering of galaxies and their velocities of mutual recession) are approximately the same in all spatial regions, if regions of sufficient size be taken, at any given cosmic instant is often known as the cosmological principle. The claim that they are the same at all cosmic instants is often known as the perfect cosmological principle. This latter principle, as we shall see in Chapter 14 (see p. 241) formed the basis of Steady State Theory. These principles

random velocities of recession of clusters from their fundamental particles virtually cancel out). For by the principle of equivalent laws the law governing the recession of the clusters will be the same relative to every fundamental particle, and so approximately the same relative to every cluster. Hence at any given cosmic instant the clusters isotropically surrounding any given cluster will be at approximately the same distances from it as those surrounding any other cluster are from it, and their velocities of recession from it be approximately the same as those of similarly positioned clusters from any other cluster. This claim is summarised by cosmologists as the claim that the Universe is always homogeneous.

The supposition that there is a cosmic time scale would have to be abandoned if new evidence suggested that the principle of similar clocks was false. Any evidence tending to show that the principle of isotropy or the principle of equivalent laws were false would tend *ceteris paribus* to show that the principle of similar clocks was false, since the arguments which I have stated currently adduced for its truth are arguments from the former principles. Other more complicated ways in which observational evidence might be relevant to the truth of these principles and the homogeneity of the Universe will be mentioned in Chapter 14.

So then between heavenly bodies within a galactic cluster, both synchronisation tests appear to give the same results — given those methods of applying them which are based on the laws of nature which result from the simplest extrapolation from observations (e.g. the law which states that light has the same one-way velocity in all directions). Between clusters the simplest extrapolation from observations indicates (via the principle of similar clocks) a reliable method of applying the clock-transport test. Given that simplicity is evidence of truth, there are

---

understood as above are undoubtedly empirical, for observations could confirm or disconfirm them.

Sometimes, however, the principles are confused with vaguer principles that the most general features of the Universe are the same in all spatial regions, if regions of sufficient size be taken, at any or all cosmic instants. The claim that all regions at any or all instants have some common properties is simply the claim that nature is uniform. The scientist looks for laws of the behaviour of matter which apply in all circumstances, and if he finds that the 'laws' which he has do not, he looks for more fundamental laws which do. Hence the scientist must of logical necessity (viz. if he is to practise science) use these more general principles.

For the history of the cosmological and perfect cosmological principles and different ways in which they have been understood, see [5] Chapter 14.

these methods of synchronisation which yield unique results which we have good grounds for believing to be true results.

I argued in Chapter 4 (see p. 68) that whether or not the distance from A to B was at some instant the same as the distance from B to A depended on our criteria for simultaneity, although if A and B are stationary relative to each other the two distances will be the same. We can now see that on the cosmic scale if A and B are clusters these distances will also be approximately the same, for if A and B are fundamental particles they will be the same. The latter is a consequence of the principle of equivalent laws. For by this principle the laws of cosmology relative to both A and B will be the same. These laws include the laws of particle recession, that a particle P will have receded from another particle Q a distance d since an instant on the clocks of each t years ago when the distance of Q from P was small or P and Q were stationary relative to each other. By the principle of similar clocks, clocks on each will record the same passage of time since that instant. Hence since the laws stating the increase of distance with time will be the same for observers on each, the increase of the distance of A from B since the original instant will be same as that of B from A. But at the original instant either the distances (of A from B or B from A) were small (relative to present distances), or they were the same (since A and B were stationary relative to each other). Hence the distances now will be virtually the same.

Since geometry has the character of a physical theory, it is a consequence of the principle of equivalent laws that the geometry of Space, as measured relative to any equibasic frame, viz. any fundamental particle, will be the same, and hence we can talk, as we saw on p. 101, of the geometry of Space *simpliciter* and not merely of the geometry of Space relative to such and such a frame. The evidence suggests that this geometry is a congruence geometry. For the supposition that the geometry of Space is a congruence geometry is simpler than any rival supposition. The only reason which scientists have had for postulating that the geometry is not a congruence geometry, is that the geometry of a region is determined by the distribution of matter in it. But if, as we have seen reason to believe, the Universe is homogeneous on the cosmic scale, then if the distribution of matter determines the geometry of a region, the geometry of all regions will be the same, and hence the geometry of Space as a whole will be a congruence geometry. Since the evidence is that the axioms of Euclidean geometry other than the fifth postulate all apply — if we are allowed to modify the axioms referred to on p. 103 which, if taken in the normal way, have to be modified to make the fifth postulate of elliptic geometry consistent with the rest of the Euclidean system —

the evidence suggests that the geometry of Space is one of the three congruence geometries discussed in Chapter 6.

It is a deductive consequence of the homogeneity of the Universe, of the applicability to it of a cosmic time scale, of the supposition that the geometry of Space is one of the three congruence geometries discussed in Chapter 6, and of the suppositions that light travels through interstellar space in straight lines and that the Special Theory of Relativity applies to a high degree of approximation on a scale smaller than the cosmic scale, that a formula known as the Robertson–Walker line element will hold. We saw earlier in this chapter the grounds for supposing that Special Theory applies to a high degree of approximation in the region of our galactic cluster. By the principle of equivalent laws it will therefore hold to a high degree of approximation in the region of any galactic cluster, and so — since it ought to hold in intergalactic regions of similar size — these being empty of matter, to an even higher degree of approximation — generally on the scale smaller than the cosmic scale. We analysed on p. 87 some of the empirical evidence showing that light travels through interstellar space in straight lines. Further, that light travels in straight lines is a hypothesis built into all such theories of mechanics as the General Theory of Relativity and so confirmed to the extent to which they are confirmed.

Now let there be two neighbouring events, E_1 occurring on a fundamental particle located by spherical polar coordinates (u, θ, ϕ) at cosmic instant t, and E_2 occurring at $(u + du, \theta + d\theta, \phi + d\phi)$ at cosmic instant $t + dt$. θ and ϕ are the normal angular coordinates. u is a 'co-moving' radial coordinate, such that if a particle has it at one instant it has it at all instants. It is related to the particle's coordinate distance r (see p. 104) by $u = r/R(t)$. $R(t)$ is a variable of time, known as the radius of the Universe, and in this way measures the extent to which particles are spread out. Then an interval ds^2 will be invariant, whichever fundamental particle we take as origin, in the following formula, the Robertson–Walker line element:[1]

$$ds^2 = c^2 dt^2 - R^2(t) \left\{ \frac{du^2}{1 - ku^2} + u^2 \, d\theta^2 + u^2 \sin^2 \theta \, d\phi^2 \right\}$$

(ds is the spatio-temporal interval between E_1 and E_2. When E_1 and E_2

---

[1] See [3] Chapter 5, for an account of the work of Robertson, Walker, and Rindler in proving the equations to be stated without proof in the next two pages.

are points on the path of a ray of light $ds = 0$.) Here c is — to a high degree of approximation — the one-way velocity of light relative to each fundamental particle through which it passes, and the mean velocity of light relative to other frames in its neighbourhood moving with uniform rectilinear velocity relative to that particle, as shown by Special Relativity which applies to a high degree of approximation in the neighbourhood of each fundamental particle. k is a constant having values $k = 0$ for Euclidean Space, $k = +1$ for elliptic space (and also for spherical space), $k = -1$ for hyperbolic space.[1] From the Robertson–Walker metric cosmologists develop various more specific 'models', that is cosmological theories, by giving a specific value to k and making some specific supposition about the equation governing the value of $R(t)$.

The adoption of the Robertson–Walker metric may commit us to a prediction that the mean velocity of light on a cosmic scale is not the same relative to all fundamental observers at all instants. The proper distance from a cluster C_a at the origin, of a photon, the particle of light, emitted from a cluster C_b having u-coordinate u_b at cosmic instant t_1 in the direction of C_a is, at cosmic instant t:

$$l = R(t)\left\{\sigma(u_b) - \int_{t_1}^{t} \frac{c\,dt}{R(t)}\right\},$$

where $\sigma(u) = \displaystyle\int_{0}^{u} \frac{du}{\sqrt{1 - ku^2}}$.

Hence the velocity of the photon at t relative to C_a, marking velocity of approach by a negative sign, will be

$$\frac{dl}{dt} = R'(t)\left\{\sigma(u_b) - \int_{t_1}^{t} \frac{c\,dt}{R(t)}\right\} - c.$$

$R'(t)$ is the rate of change of $R(t)$.

We have good observational evidence that $R'(t) \neq 0$, and so that photons emitted from different fundamental particles have at different instants different velocities relative to other fundamental particles. Hence the one-way velocity of light will vary with the fundamental particle which

---

[1] k must be distinguished from the curvature of space K defined on p. 104. For Euclidean geometry $k = K = 0$. For elliptic or spherical geometry $k = +1$ while K has a positive value which depends on how marked is the positive curvature. For hyperbolic geometry $k = -1$, while the value of K depends on how marked is the negative curvature.

forms the frame of reference. This conclusion, as we saw on p. 191, forms part of the evidence for supposing that each fundamental particle forms the most basic frame for the explanation of local phenomena and hence for concluding that the one-way velocity of light varies locally also with the frame of reference. This result can however quite easily be reconciled with the Special Theory of Relativity by representing it as claiming, as we saw that we could represent it, only that the mean or two way velocity of light is constant relative to all inertial frames while the one-way velocity varies with the frame. The above equation, however, on many natural assumptions about the form of $R(t)$ also predicts that the two-way velocity of light is not constant on a cosmic scale relative to every fundamental particle. This is to say that light sent to a distant cluster and reflected back will not in general have relative to its source a constant mean velocity c. But of course to do a Michelson–Morley experiment on a cosmic scale to test this is practically impossible. Hence we need more indirect evidence. The evidence leading up to formulation of the Robertson–Walker metric, which we have sketched, was evidence in favour of this supposition, and so to the non-applicability of the Special Theory of Relativity on the cosmic scale.

I have argued in this chapter that it is not under normal circumstances an arbitrary matter whether we say that two events are or are not simultaneous. Our ordinary criteria for simultaneity are the clock-transport method and the signal method. These need to be corrected and made precise in the ways which I have developed. Corrections are necessary because the different methods employed in different ways yield conflicting results. Which corrections we make are dictated by considerations of simplicity. The simplicity of a proposed scientific law compatible with observations is, as I urged on p. 43, evidence of its truth and so evidence of whatever judgements of simultaneity are yielded by it. I urged that our scientific evidence is such that considerations of simplicity yield unique judgements of simultaneity, a unique way of extrapolating our normal methods so as to yield non-conflicting results on the cosmic scale. However it might be that that evidence is misleading, and that in fact compatible with all actual (and possible) observations there are many different equally simple ways of expressing the laws of nature, which could be put forward and which yield very different judgements of simultaneity. We saw that this would be the case if our Universe was the Universe of Special Relativity. Here there were many equibasic frames, in terms of measurements relative to which the laws of nature could be set forward, but measurements relative to which yielded (within limits) different judgements of simultaneity. In such a case men would have no grounds for preferring one such judgement to others.

There seems however to be a difference between the case of simultaneity and the cases of measurement of spatial and temporal intervals. If there were different equally simple ways of setting out the laws of nature compatible with all actual (and possible) observations which (consonant with our first criterion of normal usage—see p. 178) yielded different judgements of temporal interval (e.g. that some interval was an hour as measured in one way, and an hour and two minutes as measured in a different way), I am inclined to say that there is no unique true judgement of how long the interval is. It is not merely that we cannot know the truth, but there is no truth to be known. For what are statements about the length of intervals except statements about what would be measured by clocks which yield measurements broadly consonant with our ordinary judgements corrected to fit the simplest ways of setting out the laws of nature? But if equally simple ways of setting out the laws of nature were to yield different judgements of simultaneity, there would seem to be a truth of which we are still ignorant. Suppose that, in the Universe of Special Relativity, a signal is sent from P to Q and reflected back to P. The clock at P ticks on from t_1 to t_5 during this interval. Surely there must be one moment during this interval at which the signal is reflected even if we can never know what that moment is. It is hard to describe in an intelligible way a universe to which the concept of absolute simultaneity has no application. And as we saw in the Introduction, if one cannot describe something intelligibly that is grounds for supposing that it is not a logically possible state of affairs. As we saw on p. 186, Einstein recommended that in the Universe of Special Relativity we abandon the concept of absolute simultaneity and substitute for it the concept of simultaneity in a frame. Einstein made this recommendation because of his strongly verificationist approach to science; but, for the reasons given in the Introduction, there is no need to follow him in this. One can describe the Universe of Special Relativity perfectly intelligibly by supposing that its equations show a limit to our knowledge of absolute simultaneity, not a limit to its existence. If that is right, the same applies generally with respect to the General Theory of Relativity. The four-dimensional-abstract entity space-time with which the theory deals (see p. 114) may be divided into three spatial dimensions and one temporal dimension in many different ways—that is, for events E_1 and E_2 separated by a certain spatio-temporal interval ds, one may make many different hypotheses compatible with evidence, as to what is the spatial distance and what is the temporal interval between them. Nevertheless whether or not there is a truth about the length of the interval between the events, there is a truth about which event occurred first, even if we cannot know it and our evidence supports equally the hypothesis that E_1 occurred first and the hypothesis that E_2 occurred first. However, as we have seen, our Universe appears to be the Universe of the Robertson–Walker line element and so one in which there are grounds for making justified judgements of simultaneity.

Newton held similar views about time, to his views about space. 'Absolute, true, and mathematical time' he wrote, 'of itself and from its own nature, flows equably without relation to anything external . . . Relative, apparent and common time, is some sensible and external (whether accurate or unequable) measure of duration by means of motion' [1]. That is, he held that there is a true time which might or might not be recorded by actual measuring instruments. I have argued above, that there are true judgements of simultaneity, and ones of which men can probably have knowledge. There are also, probably, true judgements of temporal interval between instants, although it is an empirical matter whether there are; and man can have knowledge of such judgements. In the quotation Newton seems to be claiming both that there are true judgements of simultaneity and that there are true judgements of temporal interval. So I conclude that Newton is right about time. There is in his sense Absolute Time; although there is not, in the main sense discussed in Chapter 3, Absolute Space. We saw in the last chapter that time has no metrical geometry in addition to its topology. Hence, there is no question about Absolute Time similar to our second question in Chapter 3 about Absolute Space. In Chapter 3 I argued that there could not be space without physical objects. I have already argued in Chapter 10 the opposite thesis with regard to time — there could be time when there were no physical objects.

So far I have discussed only primary tests for measuring temporal interval. They concern what would be or would have been observed by observers watching clocks and sending signals. But the Universe is not populated by observers carrying out such tests, and they and any race of them would have perished long before they could carry out some of those tests. Hence we rely on secondary tests to obtain knowledge of temporal intervals between events. We use secondary tests, it will be recalled (see p. 61) because we have good reason to believe that, had primary tests been carried out, they would have given the same results as the secondary tests.

There is not space for a long analysis of secondary tests for measuring temporal interval, similar to the analysis in Chapter 5 of some of the secondary tests for measuring distance, but two very brief examples will be given, just to illustrate that they work in the same way as the secondary tests for measuring distance by assuming that some locally tested relation between phenomena holds at other places and temporal periods than those for which it has been tested. One example is provided by the method of ascertaining the age of objects from the percentages of radioactive substances contained in them. We find from experiments at some place on Earth in the 1960s the proportion of atoms of substances which decay in some period of time, as measured by atomic clocks, viz. by primary tests, and so assuming the same rate of decay to operate at all temporal periods we find the half-life of such substances, viz. the period of time in which half of some mass of a substance will decay. We find that two isotopes of Uranium, U^{238} and U^{235} decay respectively into isotopes of lead Pb^{206} and Pb^{207} with half-lives of 4510 and 707 million years. We then infer by retrodiction the proportion of these isotopes to be expected in the Earth's rocks on their first evolving. Then from the percentage of the different isotopes of lead to the different isotopes of Uranium currently found in rocks, we calculate their age and so find the age of the oldest rocks and hence when the Earth began to condense. The same kind of method is used for estimating the age of stars from the chemical processes which must have occurred in their evolution. From stellar spectra and luminosity we can infer which chemical elements in what proportion at what temperature are to be found on the surface of stars. We retrodict the elements out of which they were originally formed. From studies on Earth we know the temporal periods, judged by atomic clocks, which various masses of various elements at various temperatures take for conversion into other elements at other temperatures. Assuming the same laws to hold for places and temporal periods for which they have not been tested, we can infer the temporal period of a star's evolution. Again, we estimate the temporal interval between events observed to occur on distant stars within our cluster (e.g. flaring up of novae) and the instant at which they are observed by supposing that light always travels at the same velocity as it is now observed to travel locally, as judged by atomic clocks.

So in order to use secondary tests to estimate the instant of occurrence of past events we assume locally tested regularities to hold far beyond the spatio-temporal range for which they have been tested. We alter our estimates of the instant in so far as we have reason to suppose that these regularities do not hold beyond this range. We have just given reason for supposing that the law governing the velocity of light travelling between clusters is different from that which, at any rate to a high degree of approximation, governs its velocity within a cluster.

One further issue concerned with time measurement must be discussed very briefly. Philosophers of science sometimes write about 'psychological time' or 'the time of our experience' and distinguish it from 'physical time'.[1] So far in this chapter I have been writing about what they would call 'physical time', that is the temporal intervals which in fact occurred between two events, and the measure of them which would be given by actual physical clocks. However, events, the temporal interval between which is in fact a certain number of minutes, often seem to observers separated by more or fewer minutes. By this we mean that had they not had a clock, they would have judged the interval differently. Because temporal intervals are not always what they seem to be, the writers distinguish temporal intervals as indicated on clocks, which they call 'physical time', from temporal intervals as we would judge them without clocks, which they call 'psychological time'. The interval of 'psychological time' between two events may then differ from the interval of 'physical time' between those events.

Now certainly temporal intervals are not always what they seem to be. But to describe this fact by introducing two kinds of time, 'physical time' and 'psychological time', is very misleading. If we are to make the point consistently in this way, then since different observers make different judgements of temporal interval if they do not have clocks, we ought to introduce as many 'psychological times' as there are observers. But the terminology is misleading because the two 'times' are not on a level. 'Physical time' describes the true time interval between two events. 'Psychological time' is not a measure of an equally correct interval between the events. It is a measure, which may or may not be correct, of how observers not using clocks, judge, as well as they can, what is the interval of 'physical time' between the events.

BIBLIOGRAPHY

[1] I. Newton, *Principia*, Scholium to Definition viii.
[2] H. Reichenbach, *The Philosophy of Space and Time* (originally published 1928) (trans. M. Reichenbach and J. Freund), New York 1958, ch. 2.

[1] See, e.g. H. Reichenbach, *The Direction of Time* (Berkeley and Los Angeles, 1966). He asks (p. 269) the question 'What is the relation between the time of physics and the time of our experience?'

[3] G. J. Whitrow, *The Natural Philosophy of Time*, London, 1961, chs 4 and 5. (This work includes a much more detailed account of the physical and mathematical basis of cosmological developments leading to the formulation of the Robertson–Walker metric than I have given.)

[4] A. Grunbaum, *Philosophical Problems of Space and Time*, London, 1964, ch. 12.

[5] J. D. North, *The Measure of the Universe*, Oxford, 1965, especially chs 14 and 16.

For details of time scales used for measuring time at any one place see:

[6] William Markowitz, article on 'Time Measurement' in *Encyclopaedia Britannica*, 1964 ed, vol. 22.

For details of the physical and mathematical basis of the Special Theory of Relativity, see:

[7] W. Rindler, *Special Relativity*, Edinburgh, 1960.

[8] D. Bohm, *The Special Theory of Relativity*, New York, 1965.

On simultaneity within an inertial frame see:

[9] Brian Ellis and Peter Bowman, 'Conventionality in Distant Simultaneity', *Philosophy of Science*, 1967, 34, 116–36.

[10] Adolf Grünbaum, Wesley C. Salmon, Bas C. van Fraassen and Allen I. Janis, 'A Panel Discussion of Simultaneity by Slow Clock-Transport in the Special and General Theories of Relativity', *Philosophy of Science*, 1969, 36, 1–81.

[11] Brian Ellis, 'On Conventionality and Simultaneity', *Australasian Journal of Philosophy*, 1971, 49, 177–203.

12 Physical Limits to Knowledge of the Universe — (i) Horizons

So far in this work we have mainly been investigating the meaning of spatial and temporal terms and of propositions about Space and Time, although in the last chapter in order to examine the application of our criteria of simultaneity, we had to set forward the evidence for certain empirical truths about the Universe. The task of the concluding chapters is to examine more fully what kind of conclusions science can hope to reach about the general spatio-temporal character of the Universe.

A necessary preliminary to this task is to inquire whether there are any necessary limits to our knowledge of the Universe at other places and temporal instants. In Chapter 9 we considered whether there are any logical limits to such knowledge. In this and the next chapter we must consider whether the laws of physics impose any limitations. Initially it might seem that, within the logical limits discussed in Chapter 9, it is a merely practical matter that we do not know about events at distant places and remote temporal instants. For any event at a distant place or remote temporal instant it is easy to imagine the constituents of the Universe being different so that we might learn about it. If only ancient papyri had not been accidentally destroyed, would we not be able to read all the plays of Sophocles? Could we not learn about events on a certain galaxy at a very remote instant if we built a radio telescope big enough to detect waves emanating from it? However, it has recently been suggested that in two important ways the laws of nature have certain characteristics which prevent our acquiring knowledge of events at distant places or remote instants. These suggestions we must consider in this and the next chapter. In this chapter we shall consider suggested limits to knowledge of states of distant galactic clusters, brought about by their fast velocity of recession from ourselves; in the next chapter we shall consider suggested limits to knowledge brought about by states being more reliable signs of past states than of future states.

The more distant a galactic cluster is from the Earth or any other heavenly body, the faster it recedes from it. Such is the well substantiated evidence of astronomy. It has been suggested for two distinct reasons that the law of recession leads to a limit to what we can detect by observation.

The first reason is that clusters may recede so fast from each other that light emitted from a cluster will never reach some clusters, since they are receding from it as fast as or faster than the light is approaching them. It is clear that if the Special Theory of Relativity were applicable on a cosmic scale this could never happen. For on this theory light has the same one-way velocity relative to all inertial frames of reference, and a cluster is to some degree of approximation an inertial frame. Hence light emitted from a distant cluster towards the Earth will have, relative to the Earth, the same one-way velocity as it has relative to the distant cluster, viz. c; and so will eventually reach the Earth, however distant the distant cluster be from the Earth. However, if we suppose, as we have done so far for reasons given in the last chapter, that it is not Special Relativity but the Robertson–Walker line element that applies to the Universe on a cosmological scale, clusters may recede from each other with the consequences described. Since we have reason to believe that no signal travels faster than light, then if light from a distant cluster never reaches a certain cluster, the former cluster will be unobservable by an observer on the latter one.

Cosmologists discuss two kinds of 'horizon' which arise from high velocity of mutual recession. A horizon is a surface in Space marking some limit to the knowledge by observation available to an observer of events occurring beyond the surface. (Horizons are sometimes depicted as surfaces in space-time, but they can be represented, and I shall represent them as surfaces in Space.) In a classic paper [1], analysing the properties of cosmological horizons, Rindler defined 'an event-horizon for a given fundamental observer A', as 'a (hyper-) surface in space-time which divides all events into two non-empty classes: those that have been, are, or will be observable by A, and those that are for ever outside A's possible powers of observation.' If we consider as we shall do in future, the horizon as a surface in space, not space-time, an event horizon will be 'for a given fundamental observer A and cosmic instant t_0 a surface in instantaneous 3-space $t = t_0$ which divides all fundamental particles into two non-empty classes: those, events occurring on which at t_0 will be observable by A at some future cosmic instant, and those, events occurring on which at t_0 will never be observable by A.' (By '3-Space' Rindler means physical Space.) Having made and used earlier the distinction between fundamental particles and galactic clusters, I shall henceforward — to make discussion simple — often ignore the difference between them and make statements about clusters which strictly speaking apply to fundamental particles, and are only approximately true of galactic clusters. Hence I shall say that an observer who remains on the Earth is a fundamental observer because he remains on his galactic cluster. The Earth's event horizon then marks a limit to knowledge by observation available to an observer who remains on the Earth.

Rindler distinguished the event-horizon from the particle horizon. He defined 'a particle horizon for any given fundamental observer A and cosmic instant t_0' as 'a surface in instantaneous 3-space $t = t_0$, which divides all fundamental particles into two non-empty classes: those that have already been observable by A at a time t_0 and those that have not.'

An event-horizon arises if any two clusters recede from each other at an ever-increasing rate, so that light emitted from one towards the other after some instant never reaches the other. Clusters lying beyond the Earth's horizon are those, events currently occurring on which will never be observable on Earth. Events which occurred on them previously may have been observed on Earth. Clusters lying beyond the Earth's particle horizon are ones which have never yet been observable on Earth. Such a horizon will arise if clusters recede from each other at an initial period of the existence of the Universe with a velocity greater than c (and also under certain other highly unlikely physical circumstances). Of the various 'models' developed from the Robertson–Walker line element considered by cosmologists as possible theories of the Universe, some have an event-horizon, some have a particle horizon and some have both. In his article Rindler showed that the necessary and sufficient condition for an event-horizon to exist at cosmic instant t_0, in a given model is, given the correctness of the Robertson–Walker line element, that the integral $\displaystyle\int_{t_0}^{\infty} \frac{dt}{R(t)}$ converge to a finite limit, where $R(t)$ is the 'radius of the Universe' at cosmic instant t. Thus the De Sitter model, where $R(t) = e^{t/T}$, and Page's model where $R(t) = at^2$ (a and T being constants) have such horizons. The necessary and sufficient conditions for a particle horizon to exist at cosmic instant t_0 is, given the Robertson–Walker line element, the convergence of the integral $\displaystyle\int_{0}^{t_0} \frac{dt}{R(t)}$ (or $\displaystyle\int_{-\infty}^{t_0} \frac{dt}{R(t)}$, where the definition of $R(t)$ extends to negatively unbounded values of t). Thus the Einstein–De Sitter model where $R(t) = at^{2/3}$ (a being a constant) has a particle horizon.[1] Models not consistent with the Robertson–Walker line element may also have particle or event-horizons. The considerations involved in saying that a model gives a correct theory of the Universe will be examined in detail in Chapter 14.[2]

The question which I wish to examine in connection with such horizons is, if they existed, to what extent would they limit our knowledge of the Universe at other places. They would be limits to knowledge by observation. The particle horizon would be a barrier only to present knowledge, not to future knowledge by observation. More fundamental, therefore, is the event-horizon. If a cluster lies beyond the Earth's event-horizon, its present and future states would be therefore, unobservable by an observer on Earth. He could, however, have observed its past states and use them as traces of its present and future states.

The grounds which the cosmologist has for taking some state of a physical object within the range of his observation as a trace of a physical object outside that range are, as we have seen, that this sort of connection has been observed locally or that the simplest scientific theory compatible with observations postulates that it will be. It is for these reasons, as we saw in the last chapter, that cosmologists infer that clusters of galaxies are distributed uniformly throughout Space and recede from each other with a velocity which is a function of their distance apart. Astronomers also find by observation, as we shall see in detail in Chapter 14, other features of the Universe which are the same in all observed

---

[1] See [1] *passim* for further details.

[2] Particles which lie now beyond A's event-horizon may previously have been observable by A. It may therefore be useful to define a further horizon, which I will call the EP horizon, as follows. The EP horizon, for any fundamental observer A and cosmic instant t_0 is a surface in instantaneous 3-space $t = t_0$ which divides all fundamental particles into two non-empty classes; those which either have been or will be observable by A, and those which never have been or will be observable by A. This type of horizon is discussed without a precise definition or name by Rindler [1] pp. 671f.

Given the Robertson–Walker metric, the necessary and sufficient condition for the existence of an EP horizon at any cosmic instant is the convergence of the integral $\displaystyle\int_{0 \text{ or } -\infty}^{\infty} \frac{dt}{R(t)}$. If an EP horizon exists at one cosmic instant, it will exist at all others. One model, which Rindler develops in detail, which has an EP horizon is the model with $k = 0$, λ positive and $R(t) = a(\cosh bt - 1)^{\frac{1}{3}}$, a and b being constants. If any cluster lies beyond both the event and particle horizons it will lie beyond the EP horizon (hence my name for it), and conversely. To obtain knowledge of clusters lying beyond the Earth's EP horizon we must observe states of other clusters, viz. ones lying within it, and uses these as traces of states of the former.

regions. For such features F, if a spatial region has that feature, then it is observed that the next region beyond it also has it. Hence F in one region is a trace of F in the region beyond it, and the latter of F in a region beyond it and so on. Thereby we can obtain knowledge of the Universe beyond the limited range, if the range is limited, which it is physically possible that we might study by observation.

Now if we had reason to believe that there was an event-horizon, a surface which could be marked in Space which was a limit to our knowledge by observation, it might appear that although knowledge could be obtained of the Universe beyond that surface, that knowledge would be shaky and unreliable and hence that that surface was the boundary to reliable knowledge of the Universe. But for two reasons the position cannot be like this. For first the very fact that we can infer — for whatever reasons — the existence of an event-horizon means that we have two important pieces of information about the physical objects lying beyond the horizon. One is that there is at least one such physical object. The other is that all physical objects beyond the event-horizon are receding from us at speeds so fast that the fastest influence which they can now transmit towards us will never reach us. These assertions follow from the definition which I gave of the event-horizon. If we have reason to know that there is an event-horizon, we have reason to know this much about objects beyond it.

But in the second place we must have used and have needed to use traces in reaching the conclusion that there was an event-horizon. The cosmologists who adopt models with horizons are only justified in doing so if these models are a justifiable extrapolation to all spatial regions of properties observed in near-by regions. We observe the distribution of galaxies and their rates of recession and other features of the Universe and from these infer the laws of cosmology, as has been shown to some extent in Chapter 11 and will be shown in more detail in Chapter 14. This enables us to take spatial regions occupied by observed clusters as traces of more distant spatial regions occupied by other clusters. Hence we learn that in all spatial regions beyond a certain distance from us there are clusters spread out homogeneously through space, all receding from each other and receding from us with such velocities that signals now sent from them could never reach us, and no physical objects of which the opposite is true. If we had reason to believe that traces were shaky and unreliable, we would not be justified in concluding that there was an event-horizon. We only have reason for believing that there is an event-horizon and so a limit to knowledge of the Universe by observation because we have knowledge by the use of traces of regions beyond observation. If the use of traces to ascertain the state of the Universe beyond the observed realm is unreliable, then our conclusion that there is an event-horizon, a limit to knowledge of the Universe by observation, is also unreliable.

However, though the existence of an event-horizon does not mean we can have no reliable knowledge of objects beyond the horizon, it does mean a limit to the type of knowledge possible about those objects. The cosmologist uses as traces of the states of unobservable regions those properties of observable regions which have proved, in so far as this can be tested, to be reliable indicators of the properties of other observed regions. These are the properties of observable regions which remain constant with the spatio-temporal distance of a region from ourselves or which vary in a systematic way with that distance. Hence the cosmologist can have knowledge of the density of matter and rate of recession of the galaxies in regions for any reason unobservable. But he cannot use as traces those properties of the regions which are neither constant nor vary in a systematic way. What those properties are, astronomers are not yet in a position to tell us. But it might turn out that, within certain limits, average cluster size or the average number of exploding supernovae per galaxy per year were neither constant nor varied in any systematic way with the spatio-temporal distance from ourselves of the regions to which they belonged. In that case we could not use traces to learn, within those limits, about average cluster size or average number of exploding supernovae per galaxy per year in regions unobserved. The fact of the existence of an event-horizon would mean that events on the Earth would be entirely unaffected by events now or subsequently occurring on objects lying beyond the event-horizon. Hence we would know nothing about average cluster size or the average number of exploding supernovae per galaxy per year in regions beyond the event-horizon. Whereas if there was no event-horizon, all physical objects could produce effects on the Earth enabling us to detect such properties. (Of course the more distant an object of given size, the more refined the astronomical instruments needed to detect any properties thereof.) So my conclusion is that the existence of an event-horizon for an observer on the Earth would not mean that such an observer could know nothing of regions beyond that horizon but that he could only have knowledge of those properties of those regions which, within the range of our observation, were constant or varied in a systematic way with spatio-temporal distance from the Earth. If it is a fact that we can only have knowledge by traces of some distant regions, that means that we can only know certain properties of those regions. Knowledge by traces which are not effects does not include knowledge of peculiarities in the way that knowledge by observation does.

The supposed existence of an event-horizon has sometimes led scientists to suggest a definition of the Universe not, as we have done, as all the physical objects that there are spatially related to ourselves, but as all the physical objects within that horizon. Bondi relates that many consider the Universe as the set of all events which could have affected us in the past and 'which may affect us at sometime in the future and all events which have been or will be affected by us'. (Bondi adds that 'some authors . . . consider a different set, viz. the largest set to which our physical laws (extrapolated in some manner or other) can be applied. . . . The physical significance of this "universe" is not very clear' ([2] p. 10).

Now while it may be reasonable to define a space-time Universe as consisting of events, the Universe of ordinary language is most naturally said to consist of physical objects. If we are to define a Universe in terms of physical objects, the definition of it best satisfying the positivistic urge behind the suggested definition would seem to be of the Universe as all the physical objects with which at any instant we could have causal links, that is, all physical objects within the event-horizon. If there were no event-horizon, all physical objects could affect and be affected by events of the Earth, and hence the Universe would be all the physical objects that there were. However, if there was an event-horizon, and we had good reason to believe that there was, we would have good reason to know about objects and their states beyond that horizon, as we have noted. Hence, on the suggested definition, if there were an event-horizon, there would be physical objects spatially related to ourselves outside the Universe about which we could claim knowledge on grounds of scientific inference — which is absurd. I conclude that the suggested definition is unacceptable, and a more reasonable definition elucidating ordinary usage is that the Universe is all the physical objects (spatially related to ourselves) that there are.

I must make one final modification. I have been discussing the limits of knowledge of distant regions of the Universe for an observer in the Earth. We saw that the event-horizon marked a limit to his knowledge by observation. However, in order to make claims about the possible knowledge of the Universe available by observation to an observer now on the Earth, we need to consider the possibility of that observer travelling in space. Hence it would be more appropriate to define the following horizon which I shall call the *ET* horizon. I define it as follows. The *ET* horizon, for any observer *A* on a fundamental particle at cosmic instant t_0, is a surface in instantaneous 3-space $t = t_0$ which divides all fundamental particles into two non-empty classes: those, events occurring on which up to t_0 have been observable by A on his fundamental particle up to t_0 or will be observable thereafter if he travels in space with some velocity possible for matter leaving his fundamental particle, and those not so observable. No events now occurring on objects lying beyond his ET horizon could ever be observed by an observer now on Earth. If we wish to consider the limitations to knowledge produced by the existence of horizons for an observer who is allowed to travel in space, it is the ET horizon that marks the limit to his knowledge of the Universe by observation. It must be noted that while an observer now on Earth may obtain knowledge by observation of events now occurring in any region within the ET horizon by travelling in space, he cannot necessarily obtain knowledge of events now occurring in all such regions — if he travels in one direction, events now occurring on some clusters situated in the other direction within the ET horizon may not ever be observable by him. The reasons for believing in the existence of an ET horizon will be similar to those for believing in the existence of an event-horizon (though some cosmological models may have an event-horizon without having an ET horizon) and the extent of limitation of knowledge is similar.

So much for the extent and kind of limit to knowledge which would be produced by the existence of horizons[1]. Such horizons will, however, only arise if photons do not have on the cosmic scale the same one-way velocity relative to all inertial frames. There is, however, another

[1] It is sometimes suggested (see e.g. [4] p. 100), that 'black holes' have event-horizons. A black hole is a large star or other massine body which has collapsed under gravitational self-attraction into a small volume so dense that all matter or radiation within a critical radius of the body cannot escape but is pulled continually closer to the centre of the body. Astrophysicists have calculated from General Relativity that the final stage in the evolution of stars having more than three times the mass of the Sun is to become black holes, and there now seems to be some observational evidence of the existence of black holes. This evidence is that of stars moving like components of a binary star system under the gravitational influence of an invisible body, losing matter which is streaming in the direction of the postulated invisible body and emitting X-rays in the process. A black hole will not be observable by detecting photons emitted from it, but it can be observed so long as it causes effects of some kind which can be detected by an instrument. (See my pp. 154f). The mass, angular momentum, and electric charge of a black hole cause effects outside the critical radius; and instruments can be constructed which provide information about the state of the black hole by measuring these effects. It is because of these effects that astronomers can locate black holes.

limit to knowledge by observation for an observer who remains on his cluster, which arises even if photons did have the same one-way velocity (c) relative to all inertial frames. This will arise because of the Heisenberg indeterminacy principle.[1] Because of this principle very little information can be received on a cluster from another cluster whose velocity of recession from it approaches c. (Given the universal application of Special Relativity, velocities of mutual recession could not equal or exceed c.) For as the velocity of recession approaches c, so the wave-length of the received light approaches $\infty \left(\dfrac{\lambda}{\lambda_0} = \dfrac{1 + V_r/c}{(1 - V_r^2/c^2)^{1/2}} \right.$, where λ is the wave-length of the received light, λ_0 of the emitted light, and so the Doppler shift $z = \dfrac{\lambda - \lambda_0}{\lambda}$, and V_r the velocity of recession). But by Heisenberg's principle, our ability to locate any particle precisely depends on using very small wave-lengths of light. As $\lambda \to \infty$ locations become confused to the point where we cannot distinguish one galaxy from another. Light coming from them too would be approaching zero intensity. (Since the energy of a photon reaching us is $E = \dfrac{hc}{\lambda}$, as $\lambda \to \infty$, $E \to 0$.) Hence no detailed observation can be made of events on clusters receding from the Earth with velocities $\to c$. But as before, we may learn about them by traces, and indeed must do so in order to affirm that there are events on clusters, which we cannot observe.

BIBLIOGRAPHY

[1] W. Rindler, 'Visual Horizons in World Models', *Monthly Notes of the Royal Astronomical Society*, 1956, 116, 662–77.

[2] H. Bondi, *Cosmology*, 2nd ed., Cambridge, 1960, part i.

[3] Carlton W. Berenda, 'On the Cosmological Principle of McCrea', *Philosophy of Science*, 1964, 31, 265–70.

[4] M. Berry, *Principles of Cosmology and Gravitation*, Cambridge, 1976.

By contrast, when galactic clusters recede from each other too fast, no effects at all are propagated from one to the other. I conclude that black holes do not have event-horizons. (For more details on black holes see [4] pp. 91–102.)
[1] I use here the argument of [3].

13 Physical Limits to Knowledge of the Universe — (ii) Past/ Future Sign Asymmetry

We considered in the last chapter the extent to which the fast velocity of mutual recession of galactic clusters limited the knowledge that we on Earth could have of their states. In this chapter we must consider how far physical laws are responsible for our much greater knowledge of past than of future.

We noted in Chapter 9 that the amount of our knowledge of the past was very much greater than our knowledge of the future, and that this was in small part a logical and in much larger part an empirical matter. If there are to be beings with knowledge they must remember and so be able to report the past stages of their arguments and must be ignorant of the kind of investigations which they will conduct into the truth of propositions, and of the results of their investigations. But given these limits, there seem no other logical factors responsible for our much greater knowledge of the past than of the future. Consequently the question arises as to the empirical source of this asymmetry. Do we just happen to be ignorant of the future in this way or is some fundamental physical law responsible?

We saw in Chapter 9 that our greater knowledge of the past is due to the fact that men can report what they have observed but not what they will observe, and that traces allowing any inference are much more frequently of the past than of the future. Now the physical source of men's ability to report straight off on what happened rather than on what will happen must lie in the characteristics of human brains since, psychologists and physiologists assure us, human memory and speech abilities depend on the constitution of the brain rather than of any other part of the body. Hence the brains of organisms must be such that their states differ in respect of different past macroscopically distinguishable events and states. Thus my brain is in some state B_1 now. It would be in a different state, say B_2, had I just bumped into you in the corridor. Whereas it will normally be in the same state B_1 whether or not I am about to bump into you in the corridor. If brains did not have this property we would not be able to report what had happened more readily than what will happen. (I neglect as scientifically implausible the possibility that brain states could be correlated equally easily with future interactions as with past interactions but that this had no effect on human ability to report about the future.) What forms these states take are currently being investigated by physiologists — the states may be chemical deposits or electrochemical rhythms.

The physical source of the asymmetry of ease of prediction and retrodiction from traces lies in the existence of traces allowing easy inference to the past rather than to the future. This means that macroscopically distinguishable events and states of objects are in human experience more reliably correlatable with past than with future macroscopically distinguishable events and states of objects. If we consider only the case where an object is a trace of its own past or future state, this means that men find that things generally begin only in one or few recognisably distinct ways but may end in many recognisably distinct ways. Men are born only in a womb as the result of fertilisation, but they die everywhere and from manifold causes. And so it is with other things. Houses are built by men, but they may be destroyed by earthquakes or floods or hurricanes or bombs. Washing machines begin their lives in a washing machine factory, but they may end them on a municipal rubbish heap or a metal scrap-yard, in a back garden or by a main road.

Now to say that there are easily recognisable traces allowing inferences to the past much more reliable than inferences to the future is to say (see p. 146) that states which men can easily recognise are, in human experience, signs of past states but not of future states. I made a distinction in Chapter 9 between the two terms which I introduced, 'traces' and 'signs'. A state A_1 is a trace of another state B_1 if men have good reason to believe it to be a sign of B_1. A_1 is a sign of B_1 if it is a member of a class of events As, the occurrence of members of which is in fact correlated to a very high degree with the occurrence of Bs.

Now to what extent is this asymmetry of ease of prediction and retrodiction a function of the signs which men can recognise easily and from which they can make ready predictions, and to what extent is it an objective feature of the world? Does it depend on the psychology and physiology of humans or would any organism however constructed find the same asymmetry?

Our ability to predict and retrodict clearly depends on our ability to recognise certain events and states. If an organism can recognise a certain state, I shall say that it has the concept of that state. Now the fact that we can infer from a footprint to a foot which made it depends in part on the fact that we have and can easily apply the relevant concepts. That is, we recognise as instances of the same concept (footprint) marks of different sizes and shapes in the sand. Many of the lower organisms cannot do this. Also we recognise as instances of the same concept ('man walking on the sand') you walking on the sand, me walking on the sand, and so on. Unless humans were like this, they could not make the relevant inferences. That they are like this depends on their brains and sense organs. So might there not be creatures with other brains and sense organs who had concepts enabling them to predict better than to retrodict? Such creatures might be able to recognise states which would subsequently normally be followed by a man walking on the sand instead of states normally preceded by a man walking on the sand. Against this I would urge that whatever the brains and sense organs of creatures the asymmetry of ease of prediction and retrodiction would very probably hold, and this because of a very general objective feature of the world, which I shall call past/future sign asymmetry. This feature is the following.

Any state of a system (that is, collection of physical objects) A_1 is normally a sign of some past state B_1 but not of any future state C_1 of that system. By this I mean that (As) are very highly correlated with some past states (Bs) but not with any future states (Cs). A state of a system is, however, normally a sign of a future state of some specifiable sub-system within it. (States of systems are very much less often signs of states of other systems than of the same system.) It is supposed throughout this description of past/future sign asymmetry that states are described in equal detail in all cases.

Thus the readings on weather instruments or the state of the ground in Hull today are signs of the state of the weather in Hull yesterday. Whereas no state of Hull today is a sign of tomorrow's weather in Hull. On the other hand the state of a much wider system today — various readings on recording instruments throughout the world — is a sign of tomorrow's weather in Hull.

A loose way of putting this point is to say that a state of an object is normally a sufficient condition of a prior state of the whole object, but only a sufficient condition of a future state of small part of the original object, not of a future state of the whole object. This is a loose way of putting the point because only if the correlation of the sign with the thing signified is invariable is the sign a sufficient condition of the thing signified. Hence our more precise formulation.

The evidence that in the sense described past/future sign asymmetry is an objective feature of the world is that more detailed scientific investigation supports our superficial impression that states and changes of state are very highly correlated with past but not with future states and changes of state of the same system. The most detailed investigation into the nature of states of systems (viz. the most detailed analysis of the nature and states of particles of systems) shows no, evidence of high correlations with the future.

It should be noted that past/future sign asymmetry exists in almost every possible region of inquiry (one possible exception will be considered on p. 227). A counter-example which is often urged is that of planetary astronomy. It is urged that the present positions and velocities of planets are as reliable signs of their future position as of their past positions. Now it is certainly true that we are able to predict the future positions of planets as easily as we are able to retrodict their past positions. But we are only able to predict future planetary positions because we assume that no large heavenly body will come near to the planets and disturb their courses in the immediate future. We make this assumption because we have observed a much larger region of the heavens than that occupied by the planets and seen that no large heavenly body is likely to intrude. We make our predictions, in other words, by observing the state of a large region of Space in the middle of which lie the planets and inferring the future state of the sub-system of the planets. One state of the large region is indeed a sign of the future state of the small region. But the present state of the region occupied by the planets is a sign of the past state of the same region. If the planetary system had been interfered with in the past, there would be evidence of that fact within the system.

Now given past/future sign asymmetry in the objective sense defined, whatever the construction of its brain and sense organs, any organism is almost bound to be able to retrodict from traces better than to predict. For there is a lot more information to be had in the world about the past than about the future, and the organism would have to have a very peculiar psychology and physiology not to benefit from it, while being able to predict. Thus to take the weather example, he would have to recognise cloud patterns as traces of future weather, but not the state of ground or vegetation as traces of past weather; to read weather-predicting but not weather-recording instruments.

A further consequence of past/future sign asymmetry is that the states of a human brain are signs of past but not of future states of that brain. Now, as we saw in Chapter 9, if a man observes something, stimuli from that thing must affect him. The evidence of physiologists and psychologists is that the stimuli affect the brain. Hence the states of a man's brain are signs of the states which he has observed but not of those which he is going to observe. Hence men are able to report what they have observed but not what they will observe.

Hence past/future sign asymmetry explains the existence of corroborated reports of the past but not of the future, as well as explaining the fact that traces allowing easy inference are much more frequently of the past than of the future. Our greater knowledge of the past than of the future is — within the logical limits set forward in Chapter 9 — due to this physical feature, past/future sign asymmetry. It is clearly a most important feature of the physical world, and the question arises whether we can trace it to the operation of any one scientific law.

There has been a well-known attempt to account for past/future sign asymmetry in terms of the second law of thermodynamics. Reichenbach [1] traced the asymmetry in detail to the operation of this law and the existence of what he termed 'branch systems'. I shall argue that Reichenbach's explanation of the asymmetry is seriously inadequate; but, before doing so, I had better set forward the second law of thermodynamics.

The first law of thermodynamics states that a body gains energy, including heat which is a form of energy, only if another body loses energy, possibly in the form of heat. If you put a cube of ice on a warm surface, the ice will melt only if the surface loses heat. The second law of thermodynamics states that energy flows from systems possessing more to systems possessing less; and so heat flows from hot bodies to cold bodies. The first law would be satisfied if the ice grew colder and the surface grew warmer; heat would have been conserved. But the second law states that this never happens — temperatures even out. The ice gets warmer and the surface grows colder. A kettle of water on the fire gets warmer and the fire gets colder. The second law gave an exact description of this observable fact by defining a quantity known as entropy (S) and stating that the entropy of a closed system always remains constant or increases, never decreases. The entropy of a closed system is the sum of the entropy of its parts in thermodynamic equilibrium (viz. at the same temperature throughout). A system is closed if it is not subject to influences from outside, and only a closed system has entropy. The entropy of a system in thermodynamic equilibrium is a function of its internal energy (kinetic and potential) and volume. The internal kinetic energy of a closed system is given by its temperature (T). Thus the entropy of a gas which has spread itself throughout a container of volume v (and hence has no potential energy) with constant specific heat for constant volume (C_v) is $S = C_v \log T + R \log v + \text{const.}$, where R is a constant for the mass and type of gas. Now suppose two equal volumes of the same gas (1 and 2) one at twice the temperature of the other. Then $S_1 = C_v \log T + R \log v + K$ (a constant) $S_2 = C_v \log 2T + R \log v + K$. Their total entropy will be $C_v (\log T + \log 2T) + 2R \log v + 2K$. Now put them close together and C_v, R, v, and K will remain constant for each volume. Any change in the entropy of the system will thus be due to a change in the temperature of the two volumes of gas. The second law states that entropy can only increase. An increase in entropy will be and can only be produced by the temperatures coming closer together. Hence the temperatures of the two gases can only get closer together. If the entropy of the total system increases to $C_v \left(2 \log \dfrac{3T}{2} \right) + 2R \log v + 2K$, the temperatures will have evened out, both volumes of gas having a temperature of $\dfrac{3T}{2}$.

In the later nineteenth century Maxwell and Boltzmann gave an explanation of the second law itself. Any system can be in any one of a large number of microstates. A description of a microstate is a description of certain parameters of all the particles of the system, these parameters being position and momentum or perhaps kinetic energy. Two microstates of the system differ if the parameters of any of the particles differ within a range — e.g. if a particle is in this small box rather than in that small box, or has a velocity in this band rather than in that band. (The size of boxes or bands which we take does not in general affect the conclusion that the second law holds — on this see [3].) Given a fixed total energy and volume of the system, there are very many possible microstates of the system. These are assumed to be equiprobable — that is, the system is at any instant, given no further information, as likely to be in any one microstate as in any other. A large number of different microstates all correspond to one single macrostate. A macrostate of the system is defined by the temperature of the different parts of the system. A system is at one instant in a different macrostate from what it is at another instant if the various parts of the system at one instant have noticeably different temperatures from the same parts at the other instant. The temperature of a part which has the same temperature throughout is defined in terms of the mean kinetic energy of the particles in the part. It can then be shown that for normal systems (e.g. volumes of gas) in our universe almost all the microstates correspond to the single macrostate of an approximately even distribution of temperature throughout the system, and so are microstates which are not macro-scopically distinguishable, and that fewer and fewer microstates correspond to macrostates of lower and lower entropy. It can be shown from this that the entropy of a system which passed through different macrostates would very probably get larger and larger until it reached maximum, at which with very slight fluctuations the system would remain.

The conclusion depends on the assumption that all microstates are equiprobable. Microstates are so defined that this assumption holds. Different types of statistics are defined by the different microstates which they distinguish and term equiprobable. The Maxwell-Boltzmann statistics assumes that individual particles can be distinguished. Hence a microstate of particle a_1 being in position p_1 with velocity v_1, while a_2 is at p_2 with velocity v_2, is a different microstate from a_2 at p_1 with v_1 and a_1 at p_2 with v_2. On the Bose-Einstein statistics particles are not dis-tinguishable, and hence the number of possible microstates is much smaller. On the Fermi-Dirac statistics no two particles can be in the same energy state, and so the number of possible microstates will be much smaller still. Again these statistics can be subdivided according to whether the equiprobable microstates are states of energy or momen-tum; and whether we assume an infinite range of possible energy or momentum states, or, following quantum theory, only a finite number. Which statistics is to be applied to a given system is a matter for empirical investigation. Thus the Fermi-Dirac statistics, applying to all particles subject to the Pauli exclusion principle, applies to electrons and protons; whereas the Bose-Einstein statistics applies to molecules having even numbers of protons, electrons, and neutrons. We must adopt that system of statistics, the macroscopic consequences of which show that its microstates are equiprobable.

In all these forms the basic law is that entropy remains constant or increases — but is now in a merely statistical form. The law now states that it is very very improbable that entropy decreases by any significant amount, but it nevertheless remains just possible that it will. Very, very occasionally a kettle of water put on a fire will freeze. The entropy graph of a closed system with initial low entropy is thus somewhat as follows:

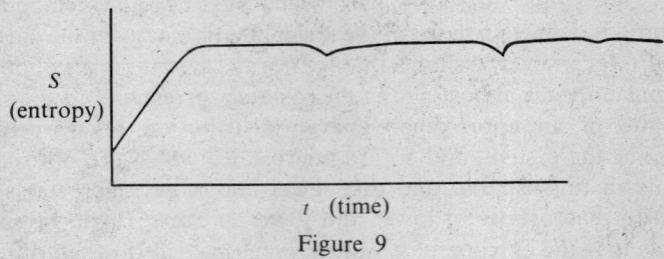

S
(entropy)

t (time)

Figure 9

For a system eternally closed (viz. having no initial state), it is just as likely that entropy will be increasing as that it will be decreasing, but far far more likely that it will be of unchanging maximum value. The frequent existence of systems of increasing entropy is to be accounted for by their not being eternally closed systems, but systems beginning in a state of low entropy.

As I have defined it and as it is defined in scientific textbooks, entropy is a function of internal energy which (if we assume that the system has no internal potential energy) means temperature, and volume. Hence whether or not we say that entropy has or has not increased depends on how small are the regions within which we are prepared to recognise a temperature difference. For suppose we divide a gas container (V) into two parts (V_1 and V_2). The entropy of the whole is the sum of the entropy of the parts. Suppose the mean kinetic energy of the molecules in V_1 is different but not greatly different from that in V_2. If we recognise this difference, we must attribute to the total system a lower entropy than if we do not. Now suppose we take smaller and smaller parts, the entropy of the parts will differ more and more, and hence the entropy of the whole become lower and lower. Any consistent application of the second law of thermodynamics presupposes general agreement on how large are the parts which we consider and how great the temperature difference between them for us to say that a system is not in thermodynamic equilibrium. Fortunately such general agreement exists on what are macroscopically significant temperature differences between parts.

Now it has become customary to illustrate the concept of entropy and indeed to give it a more general sense by introducing the notions of order and disorder, and to call a state of low entropy a state of order, and a state of high entropy a state of disorder; and to say that the second law states that, unless subject to interference, things gradually become disordered. Thus if all the molecules at one end of a container had high energies and all those at the other had low energies, the container would have different temperatures at the two ends and so be in a state of low entropy. But to have all the molecules at one end in one energy state and those at the other end in another energy state seems, intuitively, the argument goes, an ordered state — whereas to have various molecules at various places in various different energy states seems, intuitively, a disordered state. Hence the second law is sometimes said to state that disorder continually increases. The trouble with putting it this way is that there is no general agreement among people about when states are ordered and when they are not. Suppose the wind blows over a flat beach for a very long while, and eventually produces a hollow one foot deep in an otherwise flat sand surface. Is this a more ordered state than a flat beach? If it is, one of those very rare decreases in entropy of a closed system has occurred. If not, the system has remained in its state of maximum entropy. But men will clearly disagree about whether the sand with the hollow constitutes a more or a less ordered state than a flat beach. Hence no objective judgement can be made about any change of entropy of the system. It is therefore perhaps wisest to confine the concept of entropy to its thermodynamic use, where men agree about correct application.

Reichenbach used, in addition to the second law of thermodynamics, another fact in his explanation of past/future sign asymmetry. This was the existence of 'branch systems', a large number of which are at some instant in a state of medium or low entropy. A branch system is a system temporarily isolated from interaction with its environment. A clear thermodynamic and so non-controversial example of a branch system having at some instant a state of medium or low entropy would be a block of frozen ice-cream and some hot chocolate sauce in a vacuum flask.

Now, Reichenbach argues,[1] given the second law of thermodynamics and the existence of branch systems, many of which begin in a state of low entropy, past/future sign asymmetry becomes explicable. Many of these branch systems are in a state of medium or low entropy. Now they could not (or could only very rarely) have reached the medium or low state from a high state — this would violate the second law; nor could they have remained in the medium or low state for a long period for the same reason. They could only have reached the medium state by evolving from a state of low entropy, and they could not have remained for a long past period in a state of low entropy. Hence the system must have become isolated in a state of low entropy from a larger system. The

[1] I summarise the argument of §§ 15–18 of [1].

entropy of the large system as a whole would have increased while the entropy of a part decreased, after which the part became isolated. Many large systems are so constructed as to have entropy decrease in one part at the expense of it increasing elsewhere. Thus the system of the Sun and the Earth produces decrease of entropy in plants and animals who use heat from the Sun in building up large organic molecules, while the entropy of the whole system increases as the Sun's energy is utilised in the other parts. A refrigerator is another example of such a system. So if a small system is at present isolated, Reichenbach argues, the fact that it has medium or low entropy is a guarantee that it was once part of a large system and has become isolated from it. Hence its present state will be a sign of a past interaction. The states of systems possessing high entropy are not, however, signs of past interactions, for the systems may have remained isolated for an indefinitely long period.

There are in contrast no systems, the states of which are signs of their future states or changes of state. For, although we know of any system that if it remains isolated from the environment it will proceed to a state of maximum entropy, its present state is never a sign that it will remain isolated from its environment and its present state allows us to say nothing of its future state unless it remains isolated from its environment. So, the argument goes, the second law of thermodynamics and the existence of many branch systems at some instant in a state of medium or low entropy accounts for the existence of past/future sign asymmetry.

Now this argument with two qualifications which I must make seems reasonable for the ice-cream and chocolate sauce example. If there is cold ice-cream and hot chocolate sauce in a vacuum flask, it follows from the second law that the differences of temperature of the two items are a sign of a recent event of this system being isolated. But — and this is the first qualification — the second law does not inform us in what state of entropy lower than its present state the system was isolated. Perhaps the ice-cream was very cold and the sauce very hot when put in the flask, or perhaps they were much nearer in temperature. The mere fact that the second law is operating provides no guarantee that the present state of the system is a sign of any particular state of entropy of the system at the instant at which it was isolated. Secondly, any system can be in a certain state of low entropy, as opposed to a state of maximum entropy, in various macroscopically distinguishable ways. Given the present moderately hot temperature of the chocolate sauce and cold temperature of the ice-cream, the low entropy may have been realised by all the chocolate sauce being at one high temperature and all the ice-cream at one low temperature. Alternatively the ice-cream may have come from two different blocks at very different low temperatures and the chocolate sauce from two different saucepans at very different high temperatures. The two different dollops of ice-cream and chocolate sauce may have been so put together that the two pieces of ice-cream reached a common temperature by conduction, and the two splodges of chocolate sauce also reached a common temperature by conduction. These macroscopically distinct states in which the ice-cream and chocolate sauce may have been put into the flask, have the same entropy value. By contrast, maximum entropy can only be realised in one macroscopically distinct way. The lower its entropy value the more macroscopically distinct forms a system can have. These two qualifications mean that although the operation of the second law guarantees that a system in a state of medium or low entropy is a sign of an interaction in the not too distant past, it does not guarantee that it is a sign of any particular interaction. The state of medium or low entropy is not, as such, an instance of a kind of state highly correlated with a particular kind of past state. Hence we must conclude that while the second law and the existence of branch systems may be sometimes in part responsible for the existence of past/future sign asymmetry, they are not ever wholly responsible for the fact that some state signifies a prior but not a posterior state or event. For there must be some other general characteristic of the Universe in virtue of which states are signs of particular past events and states.

The case is however much worse with most examples of signs allowing easy inference to the past, including those normally cited such as footprints or fossils or relics. For on the narrow definition of entropy, these are not systems of low entropy at all. True, the entropy of a stretch of sand with a footprint is very slightly lower than that of a flat stretch of sand, since in the former case the system has slightly greater internal potential energy (work could be done by the grains of sand on the edge falling into the footprint). But the entropy difference is small, and further, the entropy of a stretch of sand with a footprint and a stretch of sand with a small hollow in it are, on the strict definition of entropy, the same. Yet the proponents of this theory, such as Reichenbach, would not wish to describe the sand with the small hollow as, in their wider sense, a system of low entropy. Why then do they wish to describe the sand with the footprint in it as a system of low entropy? It is because, according to Reichenbach, it is a 'highly ordered state, a state which in the history of the system is very improbable' ([1], p. 150). But all states, if defined by the positions of the grains of sand (viz. defined as microstates) are 'highly improbable', and yet we would not for that reason call them ordered. Even if states are classed by obvious macroscopic pattern, a stretch of sand with a small hollow in it is as recognisable as a stretch of sand with a footprint in it. Quite obviously why we want to call the sand with the footprint an ordered state is because it is the kind of state the occurrence of which would be very improbable unless it had been produced by a human foot. It is the fact that we believe that the state is a sign of a past or future state that makes us call it an ordered state. To say that the state is ordered, or, on this wide understanding of entropy, is a state of low entropy, is just to say in other words that it is a sign of a past or future state or event. Consequently entropy does not provide a physical explanation of the existence of the sign.

The same applies to all other states of a system (not being states of low entropy on the strict definition) which are signs of its past interactions, as also to states of a system which are signs of its future interactions — if there were to be any such — said to be ordered states of systems, and hence, on this wide understanding of entropy, to be in a state of low entropy. Take the case of fossils defined simply as patterns in rocks resembling in shape living organisms. Why do we consider that a rock with a fossil in it is in an ordered state? Because we believe that fossils are relics of long-dead organisms; that the rock would not have a fossil in it had not some long-dead organism been buried in it. It is termed ordered simply because it is the sign of a prior interaction. Before the latter part of the eighteenth century, most poeple believed that fossils orginated spontaneously in rocks, that they were not signs of prior interactions. Holding this belief, they would not have classified fossils as ordered states.

It follows that the fact that fossils and footprints are signs of past rather than future cannot be attributed to their being closed systems in states of medium entropy subject to the second law of thermodynamics. For on the narrow definition of entropy they are not states of medium entropy, and they are only states of medium entropy on the wide definition because we define them as being so in virtue of their being signs of past or future events or states of the system. Hence we must conclude that the second law and the existence of branch systems does not ever provide a full explanation and normally does not provide even part of the explanation of the existence of past/future sign asymmetry.

I do not know of any other single law of nature or characteristic of many laws which can in any large measure account for the existence of past/future sign asymmetry.[1] It is a task for the science of the future to

---

[1] Quantum Theory cannot account for past/future sign asymmetry. Although, according to it, there is on the very small scale a limit to the accuracy of prediction

discover the law or separate out the peculiar feature of scientific laws which is responsible for this very general characteristic of the world.

When we use some state of the Universe at one place at one instant as a trace of its state at some other place at some other instant, we are only justified in extrapolating in this way to the extent to which local evidence shows that extrapolation is reliable. Local evidence is that while states of a system are signs of past states of that system, they are only signs of future states of a sub-system. Traces are states believed with reason to be signs. Hence I am only justified in taking a state of a system as a reliable trace of a future state of a sub-system within it. This is why when the scientist predicts the future state of a whole system from its present state, he adds the proviso that the prediction is only reliable on the supposition that the system remains closed, viz. subject to no outside influences. Scientific retrodictions, however, contain no such limiting clause — no geologist infers the state of rocks a million years ago, subject to the system not having been interfered with.

Consequences for cosmology follow directly from these considerations. If we know about the present state of the Universe in a region R, we shall be able to infer with high reliability from traces, if we are able to identify them correctly, to the past state of the Universe in R, but only to the future state of the Universe in a region smaller than R. This is because physical objects now outside R may subsequently invade R or influence the behaviour of objects in R without there being any present indication in R that this will happen, but past interactions will have left a present mark. The only exception to this arises if R is a closed region, viz. is the whole of a finite Universe. For then there will be no outside influences to disturb the behaviour of objects in R. Here alone prediction could become as easy as retrodiction.

Since in this chapter we have been discussing the second law of thermodynamics it will be appropriate to refer here to another feature of it and of some other laws and processes which is often mentioned in discussion of its relation to time.

The second law of thermodynamics is often cited as an irreversible law. This term has been used in two entirely different senses and these I must distinguish clearly. I shall distinguish between laws which describe A-irreversible processes and laws which describe B-irreversible processes. A

---

from present to future states; this limit is not detectable on the macroscopic scale. Anyway it is doubtful whether Quantum Theory imposes greater limits on the accuracy of predictions than on the accuracy of retrodictions (for this, see ch. v).

process of a closed system passing from a state S_1 on a certain path to a state S_2 is A-reversible if, given the occurrence of S_1 and the rate of change of state at the instant of the occurrence of S_1, the passage from S_1 to S_2 is as probable as the passage along the reverse path from S_2 to S_1, given the occurrence of S_2 and that the rate of change of state at the instant of the occurrence of S_2 is the opposite of that on the original passage. If the laws are deterministic and, given S_1 and a certain rate of change at the instant of the occurrence of S_1, the system inevitably passes to a state S_2, then if the process is A-reversible, the reverse passage from S_2 to S_1 will be equally inevitable, given S_2 and the opposite rate of change at the instant of the occurrence of S_2 to that on the original passage. Another way of putting the claim that a law describes an A-reversible process is to say that the law accounts equally well for the succession of states if we substitute $-t$ for $+t$ in the equations describing the states. The same law would account for the process apparently depicted by a film of the original process played in reverse. The law will be A-irreversible if one passage is more probable to occur than the reverse passage, given S_1 and the rate of change at the instant of its occurrence for the first passage, and for the reverse passage S_2 and the opposite rate of change of state of S_2 at the instant of its occurrence to that at the instant of its occurrence on the first passage.

All the laws of classical physics discovered before the second law of thermodynamics are A-reversible laws. Thus if we have a state S_1 of four particles at distances each of one foot from a central point, and S_2 those four particles at distances each of two feet from the central point along the same radii as before, then the law of inertia allows equally passage from S_1 to S_2, and from S_2 to S_1. Given certain velocities of the particle at the instant of the occurrence of S_1, the system inevitably passes from S_1 to S_2. If we reverse the velocities of the particles at S_2, the reverse passage of the system is equally inevitable.

Now many changes which we meet in chemistry and biology are apparently A-irreversible — marigold seeds turn into marigolds but a film of the process in reverse would not appear to depict a physically realisable process. The first A-irreversible law to be studied by science was the second law of thermodynamics, and it is to the A-irreversibility of this law that the A-irreversibility of many of the changes of chemistry and biology referred to have been attributed. In its original pre-statistical form the second law was A-irreversible because it stated that increase of entropy could occur but decrease could not. In its statistical form the second law is A-irreversible because it states that it is more probable that a closed system in a state of increasing low entropy should pass along some path to a state of high entropy than that a system in a state of high entropy beginning to travel along the reverse path should finish its journey along the path to the state of low entropy.

However, the second law of thermodynamics is not a basic law of nature, in the sense of one that holds for any arrangement of matter and energy. It holds only for those arrangements of matter and energy in which all microstates are equally probable (i.e. equally likely to occur in the long run), when almost all microstates correspond to the single macrostate of an approximately even distribution of temperature throughout the system. Although as we have seen (p. 221) microstates are so defined that the former holds, whether the latter holds is a matter which depends on the positions and other parameters (e.g. velocities) of bodies at some moment throughout the system. The second law would not hold if matter were so arranged that (given the operations of the other laws of nature) the most frequent microstates were those which correspond to an uneven distribution of temperature. It would not hold, for instance, in a container in which all the molecules of a gas moved parallel to the walls of the container with equal velocity and in which they were all together at some instant at one end of the container. Under those conditions the basic laws of nature would ensure that the uneven distribution remained. Our Universe, or at any rate our spatio-temporal region of it, is one in which matter and energy are in fact so arranged that the second law holds.

Physicists have continued the search to see if any basic law of nature is A-irreversible. The most hopeful subject for study has been the laws governing particle decay. Very many different processes of particle decay have been investigated, but only one has been found which is not A-reversible. This is the decay of the neutral K meson into a positive and a negative pi meson ($K^0 \rightarrow \pi^+ + \pi^-$). Scientists have good reason to believe that all sub-atomic changes are PCT-invariant; viz. that the process obtained by reversing P(parity), C(charge) and T(time) in any physical process is itself an equally probable physical process. It was discovered in 1957 that K meson decay violated CP-invariance (viz. that the process obtained by reversing both C and P was not equally probable). It followed that the process would only be PCT-invariant if it was not T-invariant. In 1968 more direct experimental evidence was obtained that the process was not T-invariant.[1] A film of the process in reverse would not depict an equally probable physical process. But this

[1] For the original papers see R. S. Casella *Physical Review Letters*, 1968, 21, 1128–1131, and 1969, 22, 534–6.

law is an isolated case, and seems unlikely to have many effects on a macroscopic scale. In general the basic laws of nature seem to be A-reversible.

I distinguish laws describing A-irreversible processes from laws describing B-irreversible processes. A process of passage between a state S_1 and a state S_2 is, on my definition, B-reversible if the occurrence of passage from S_1 to S_2 is as probable as the occurrence of passage from S_2 to S_1, and B-irreversible if this condition does not hold. The difference from A-reversible processes is that I do not write here 'given the occurrence of S_1 and the initial rate of change of state' or 'given the occurrence of S_2 and its rate of change being the opposite of that on the original passage'. Now on this definition, passage from a state of high entropy to one of low entropy is not a B-irreversible process. This is because, as I noted on p. 222, in a system closed for an infinite period, passage from a state of high entropy to one of low entropy is just as probable an occurrence as the reverse passage. But the second law of thermodynamics describes an A-irreversible process because if one system is (as is highly improbable) in a state of low entropy, it is very likely to pass quickly to a state of high entropy, whereas if (as is highly probable) the system is in a state of high entropy, it is very unlikely to pass quickly to a state of low entropy. Although it is highly probable that once you have them, marigold seeds will turn into marigolds and highly improbable that once you have them, marigolds will turn into marigold seeds, a state of an eternal Universe in which one of the processes is occurring is just as likely as one in which the other process is occurring.

However there are B-irreversible processes in nature. Examples recently discussed are the examples of light waves and water waves.[1] Suppose a source of light O to send photons radially in all directions until they illuminate the walls of a container surrounding the source. S_1 is the state of the system when the source emits photons; S_2 its state as the walls are illuminated. Now the passage from S_1 to S_2 is an A-reversible sequence. You can push photons back from widely separated positions to a central one just as you can push billiard balls back. But the occurrence of the passage S_2 to S_1 is highly improbable. This is because, unless the container is spherical with internal reflecting walls, the only way to initiate the reverse passage is for a clever lighting engineer to rig up the container so that its walls are light sources emitting photons at such different instants as to illuminate O simultaneously. The existence

---

[1] Attention was called to B-irreversible processes in my sense of the term, by K. R. Popper in a series of articles in *Nature* [2].

of such an apparatus is vastly improbable compared with the existence of light sources, and hence the process of a spreading wave vastly more probable than that of a contracting wave. If the container is spherical and the internal surfaces reflect, contracting waves may be produced by reflection of the original wave, but containers seldom are spherical with reflecting surfaces. So too for other waves — water waves and sound waves spread outwards, they do not ordinarily converge. Hence we have here *B*-irreversible processes.

The limitation that the reverse process may occur if either the original process has been reflected or an agent has constructed a clever apparatus can be removed by considering processes of temporally infinite duration.[1] Thus consider a source of light, the waves of which spread out for ever into infinite space (except in so far as impeded by a relatively small number of solid bodies). The light waves from a star, if space is infinite, will spread further apart for ever. The reverse process to this process appears never to occur — light waves have never within human experience converged to a point from infinity. Clearly no agent could make them do so — for he could not rig up the machinery to effect the convergence at any instant of time at all, for any such instant would be too late; since the waves are to converge from infinity, they ought already to have been on the way. Nor could they converge as a result of an original process of divergence, for the original process would never have been finished. Yet convergence from infinity is ruled out by no law of nature. Some events are brought about by motions from infinity — as Popper [2] notes, a comet moving on a hyperbolic path around the sun may have moved from infinite time on that path. Yet what does not happen as a result of motion from infinity is coalescence, many things coming together. That such things do not happen is an important general characteristic of our universe. They are not ruled out by the way in which objects change their characteristics, viz. by physical laws, but by the distribution pattern of objects at any instant. The processes are not physically impossible; they just do not happen. They could happen given the requisite initial state, but the requisite initial state never has occurred and never will occur. Hence the process of light waves spreading into Space, would be, if Space is infinite, *B*-irreversible in the very strong sense that the reverse passage never occurs.

Now the existence and operation of irreversible processes is an interesting physical characteristic of our Universe. But some writers on physics express this interesting characteristic in a highly misleading way.

[1] See the article by E. L. Hill and A. Grunbaum [2].

They write of such processes showing 'the direction of time' or 'time's arrow'; and the suggestion is therefore that, if these processes occurred in reverse (e.g. entropy decreased instead of increasing), time would 'go in the opposite direction'. But what does this mean? That the future would precede the past, instead of following it? This however, as we saw on p. 132, is logically impossible in view of what we mean by 'past' and 'future'. All that these writers in fact mean by such flowery phrases is that the processes in question would occur in reverse.[1]

There is no logical necessity for any of the irreversible processes described in the preceding pages to have the temporal direction that they do. We described in Chapter 9 the criteria for establishing what has happened and what will happen, and none of these presupposes that any of the irreversible processes described cannot occur in the reverse temporal direction. Men might, it is logically possible, observe that kettles put on fires gradually freeze, etc. Certainly in such a universe men might not have a very long life, but that is not the point. The point is that the necessary irreversibility of the processes described in the last few pages is a physical and not a logical one.

BIBLIOGRAPHY

[1] H. Reichenbach, *The Direction of Time* (ed. Maria Reichenbach) Berkeley and Los Angeles 1956.
[2] Articles in *Nature* by K. R. Popper (1956, 177, 538), R. Schlegel (1956, 178, 381) and reply by K. R. Popper (1956, 178, 382), E. L. Hill and A. Grunbaum (1957, 179, 1296) and reply by K. R. Popper (1957, 179, 1297), R. C. L. Bosworth (1958, 181, 402) and reply by K. R. Popper (1958, 181, 402).
[3] A. Grunbaum, *Philosophical Problems of Space and Time*, 2nd edn, Dordrecht, Holland, 1973, ch. 19.

---

[1] Thus Reichenbach (ch [1], p. 139) supposed that there might be 'a galaxy in which time goes in a direction opposite to that of our galaxy'. But he only supposed this because he had defined ([1], p. 127) 'the direction of time' as 'the direction in which most thermodynamical processes occur'.

14 The Size and Geometry of the Universe

In Chapter 6 we considered what it meant to say of Space that it had a certain geometry, was finite or infinite and, if finite, had a certain volume. In Chapter 9 and in the last two chapters we have examined the amount of knowledge we can have about states of objects at other places and temporal instants and the degree of its reliability. We are now in a position to examine how we can reach a conclusion about the size and geometry of our Universe.

We should understand by the Universe, as was explained in Chapter 1, all the physical objects that there are spatially related to the Earth. The concept of the size of the Universe is ambiguous. We may understand by it the size of the space in which those objects are situated. I shall call this the s-size of the Universe. But we could understand by the size of the Universe the size of the smallest volume of space which enclosed all the physical objects that there were. (Since the constituents of the Universe are separated by empty space, to make this concept precise, some convention would have to be laid down as to how the enclosing surface was to be drawn round the edge of the outermost objects.) I shall call the latter concept the m-size of the Universe. The difference between the two concepts can be easily seen. From the seventeenth century onward men believed without question that the s-size of the Universe was infinite. However far you went along a straight line, you could always go on, coming to new regions of space getting all the time further from your starting-point. But they argued much about whether the visible Universe of stars and planets formed an island in an otherwise empty space or whether, however far you went along any straight line, you would come to new stars and planets. Their argument was about the m-size of the Universe. To say that the Universe is s-finite or s-infinite is to say that the space in which its constituents are situated is finite or infinite. The meaning of the latter statements was discussed in Chapter 6. To say that the Universe is m-finite is to say that the smallest volume containing all the constituents of the Universe is finite. To say that the Universe is m-infinite is to say that no finite volume can contain all its constituents. If the Universe is *s*-finite, then — of logical necessity — it is *m*-finite, and if it is *m*-infinite, then — of logical necessity — it is *s*-infinite, but the converse relations do not hold. Finally, I understand by the geometry of the Universe the geometry of the space in which the objects of the Universe are situated. This seems the clear meaning of an unambiguous expression.

We discussed in Chapter 6 arguments to show that Space must of logical necessity have a certain geometry or be infinite, and showed their worthlessness. The *s*-size and geometry of the Universe is a matter for empirical investigation. Arguments to prove that of logical necessity the Universe is *m*-finite or *m*-infinite are equally worthless. Such arguments on these issues I shall term *a priori* arguments.

Kant [1] put forward what he considered to be two very good *a priori* arguments, one to show that the Universe was *m*-finite and the other to show that it was *m*-infinite. The argument that the Universe is finite (viz. *m*-finite) is set forward as Kant's thesis. In order to conceive of the Universe being infinite, he argues, we should have to conceive of an infinite number of parts of the Universe, viz. physical objects composing it, having been enumerated. But to conceive of this we should have to imagine ourselves having counted the parts of the Universe for an infinite time. 'This, however, is impossible. An infinite aggregate of actual things cannot therefore be viewed as a given whole, nor, consequently, as simultaneously given' ([1], B.456) and hence the Universe must be finite. But in answer to Kant we may say that certainly, we cannot picture in our mind's eye an infinite Universe, but we can understand what is meant by saying that there is no limit to the number of physical objects of finite volume which can be counted; and in the way to be outlined in this chapter, we can coherently describe evidence which would support this claim. Hence the alternative to the finite Universe is not inconceivable. The argument that the Universe is infinite is set forward as Kant's antithesis. To suppose that the Universe is finite, he urges, is to suppose a finite world surrounded by empty space. But empty space is not a thing. So we suppose a relation between something and nothing — which is meaningless. So, he argues, a finite Universe is inconceivable, and hence the Universe must be infinite. But we have only to phrase the claim that the Universe is finite a little more clearly to avoid this objection. To say that the Universe is *m*-finite is to say that there are no physical objects spatially related to the Earth, except those within a certain finite distance from the Earth. In this form of the proposition we are not relating something to nothing. I conclude that neither the argument of Kant's thesis nor that of his antithesis is valid.[1]

Kant claimed that, given that talk of the Universe as a whole was proper, both his conclusions were correct. Since, however, the conclusions contradicted each other, this only showed that all talk about the Universe as a whole was improper. He urged a similar conclusion from the alleged fact (to be discussed in our next chapter) that he could prove both the finitude and infinitude of the age of the Universe. We shall argue that Kant's arguments on that topic also are bad arguments. Kant has not substantiated his claim that all talk of the Universe as a whole is improper. He put forward his claim that an antinomy (viz. contradictory conclusions) arose if we talked of the Universe as a whole, to substantiate his general claim that any talk of objects which could not be 'given adequately in experience' was improper. The Universe, he claimed, was such an object, because it consisted, not of a certain number of (to use his term) 'phenomena', but of all the phenomena that there are. If taken seriously Kant would seem tt be claiming that we cannot make proper assertions about all members of a class, all the so-and-sos that there are. But quite obviously we can talk of all swans, or all pieces of iron — even though we cannot picture in our mind's eye the totality of these. If a man claims that, although all talk about all swans and all pieces of iron is legitimate, nevertheless talk about all physical objects spatially related to the Earth is not, the onus is on him who makes this claim to prove it, since the logical status of the two kinds of talk is very similar. Scientists discuss coherently whether all physical objects (meaning physical objects spatially related to the Earth) have mass or have a minimum volume. Why should they not discuss whether there is a limit to the number of such objects?

I conclude in the absence of counter-arguments that talk about the Universe as a whole is proper but that arguments to prove that the Universe of logical necessity is *m*- or *s*-finite or infinite or has some specified geometry fail. The size and geometry of the Universe is an empirical matter. Our Universe is clearly very large and we on Earth can only observe a very small portion of it. Hence we must use observed states as traces of unobserved states and thereby, if we can do so, learn about all the unobserved objects that there are and so reach a conclusion

[1] Aristotle used arguments of a mixed logical and empirical character to show that the Universe was *s*-finite and so *m*-finite which are of mere historical interest: *On the Heavens*, 276a–279b.

about the size and geometry of the Universe. The scientific cosmologist argues from such evidence about parts of the Universe distant in space and time as he can collect (within the limits, logical, and physical, discussed in Chapters 9, 12, and 13) to the structure of the whole. Such arguments from particular features of the Universe to its general character I will term *a posteriori* arguments. The procedure is to use traces to infer the density of matter and the geometry of Space and other features in regions immediately beyond observation and to use these traces to infer the density of matter and the geometry of Space in more distant regions, and so *ad infinitum*. Thus we may ascertain the metrical geometry of Space as a whole and the density of matter throughout the Universe. The geometry of Space may prove to be such that if we take a certain finite number of spatial regions of finite size beyond those which we observe, we eventually return to our starting-point. Then these will be all the regions that there are. As we saw in Chapter 6, if we examine all of many observed regions and find that the geometry is elliptic of uniform curvature, and so we use each region of a certain finite size with that geometry as a trace that beyond it there is a region with the same geometry, we should have to conclude there were only so many regions of that finite size. The evidence from traces is then that Space is finite, that is, the Universe is s-finite. The evidence from traces may, however, be that however many spatial regions of finite size we consider, there is always another one. The evidence is then that Space is infinite, that is the Universe is s-infinite. If the evidence from traces is that there is no limit to the number of regions of finite size occupied by matter, then it shows that the Universe is m-infinite. If the evidence is that there is a limit, then the evidence is that the Universe is m-finite. If we were to observe that in each examined region of Space of volume V at some distance d from the Earth the quantity of matter was $k - d + |k - d|$ ($|k - d|$ means the positive value of the difference between k and d), k being a constant, then the evidence is that after a distance $k = d$ from the Earth there is no matter and hence that the Universe is m-finite. The evidence, however, which we have sketched in Chapter 11 is that the Universe is homogeneous, that in each region of sufficient size there is the same volume of matter. Hence we have reason to believe that if the Universe is s-infinite, it is m-infinite, and that if it is m-finite, it is s-finite.

The question therefore is whether the Universe is s-infinite, whether Space is infinite. To answer this question we must ascertain the geometry of distant regions and so see whether there is a limit to the number of regions of finite size. To do this we need a cosmological theory showing how the geometry of Space varies, if it does, from region to region, whether dependent on or independent of the density and distribution of physical objects. With its aid we can use the states of observed regions as traces of the states of regions beyond them.

The cosmologist endeavours to put forward a scientific theory which will describe and explain the overall behaviour of the Universe, the interactions and distribution of its constitutents on the large scale. His data fall into two groups. One group is the results of observations and experiments near the surface of the Earth explained at a lower level by such physical theories as the electromagnetic theory and the Special Theory of Relativity. The other group is the results of telescopic observation into parts of the Universe distant in space and time from ourselves. The cosmologist endeavours to build a coherent theory of the Universe from which the observed data can be derived and new predictions made which can be tested.

Most modern cosmological theories derive from the General Theory of Relativity. This was formulated by Einstein in 1916 [2] solely on the basis of the first class of data. Einstein's Special Theory of Relativity had shown, as we saw in Chapter 3, that the laws of physics referred to any inertial system were the same. Einstein felt that it was somehow arbitrary to refer the laws of nature to inertial systems; and he sought, as we noted, to set them forward in a way so that they could be referred to any *E*-frame of reference whatsover. They should hold just as truly on a frame rotating relative to an inertial frame as on an inertial frame. The requirement that the laws of nature have the same form when referred to any coordinate system (that is, any *E*-frame) whatever is known as the principle of covariance. Further, it seemed a weakness of previous mechanics that the accelerated motion of a body under the force of gravity was described as motion under a force. Since all bodies are equally subject to gravity, why not describe it as a natural motion (viz. one not due to the operation of a force)? Einstein's principle of equivalence states that to describe the effect of a gravitational field is equivalent to referring the motion of the body to an accelerated frame of reference. Since by the principle of covariance the laws of nature are to be the same in all coordinate frames, the body's motion is thus a natural one. Aiming to generalise mechanics and to some extent optics in these ways, Einstein set forward what he considered to be the simplest most general equations of motion. These were that free particles move along what he called the geodesics of space-time. (Space-time is our three-dimensional space with a fourth dimension added for time. The use of this four-dimensional geometry is, as noted in Chapter 7, a mere convenient calculating device. Geodesics are paths in space-time analogous to the straight lines of ordinary physical space. A special kind of geodesic is a null geodesic, along which light must pass.) Einstein then set forward a formula relating the geometry of space-time to the distribution of matter which would make possible calculation of geodesic paths. This formula was chosen as the most simple possible form of such a principle compatible with the data of mechanics and optics available to date. The formula summarising in tensor notation ten equations, known as the field equations, is as follows:

$$R_{\mu\nu} - \tfrac{1}{2} R g_{\mu\nu} + \lambda g_{\mu\nu} = -\frac{8\pi G}{c^2} T_{\mu\nu}$$

In this formula G is the constant of gravitation, and c the one-way velocity of light. (Einstein, as we have seen, in formulating the Special Theory of Relativity, believed the latter to be the same in all inertial frames. If we do not hold this, c should be defined as the mean velocity of light relative to the fundamental particle through which it is passing.) $T_{\mu\nu}$ is the energy-momentum tensor determined by the distribution of matter and radiation. All terms on the left-hand side of the formula are ultimately composed, as well as of the constant λ, of the metric tensor $g_{\mu\nu}$ and its derivatives. $g_{\mu\nu}$ describes the geometry of space-time and hence for any given cosmic instant the geometry of space. It measures the curvature of Space. Einstein's formula thus shows how geometry is determined by the presence of physical objects.

On a local scale there have been a number of tests of the General Theory, which confirm it highly. Two phenomena which Einstein classically adduced as evidence in favour of his theory were the advance of the perihelion of Mercury, and the deflection of light rays near the surface of the Sun. In both cases General Relativity predicts much larger effects than those predicted by Newtonian mechanics, and in both cases the predictions of General Theory are very close indeed to the best values obtainable by current observation. In the last fifteen years there have been a number of further successful tests of General Relativity. But it should be noted that they are mostly tests of what is known as the Schwarzschild solution of General Relativity, i.e. of the differences to the behaviour of material objects and radiation which are brought about by the presence of one large massive body (such as the Sun or Earth) in the neighbourhood; and they are all tests which concern the behaviour of material objects and radiation within our solar system.

To apply the General Theory on the cosmic scale we have to feed into its equations astronomical data. The most important of these are the data described in Chapter 11 leading up to the claim that the Robertson–Walker line element is applicable to the Universe on the cosmic scale. If we assume this claim, the equations of General Relativity reduce to the following relatively simple formulae:

$$\frac{8\pi G}{c^4} p = \lambda - \left\{ \frac{k}{R(t)^2} + \frac{R'(t)^2}{c^2 R(t)^2} + \frac{2R''(t)}{c^2 R(t)} \right\}$$

$$-\frac{8\pi G}{c^2} \rho = \lambda - \left\{ \frac{3k}{R(t)^2} + \frac{3R'(t)^2}{c^2 R(t)^2} \right\}$$

k and $R(t)$ have the role described on pp. 198f. $R'(t)$ is the first derivative of $R(t)$ with respect to time (t) (viz. it measures the rate of change of $R(t)$ over time) and so is a measure of the rate of mutual recession of the clusters. $R''(t)$ is the second derivative of $R(t)$ with respect to t (viz. measures the rate of change of $R'(t)$ over time). p is the mean internal pressure of matter (and radiation), and ρ its mean density. G is the constant of gravitation, and c the velocity of light, as before. λ is the cosmological constant, whose value is only calculable through the field equations. Thus G and c are known constants; k and λ unknown constants; p, ρ, $R(t)$, $R'(t)$ and $R''(t)$ are time-dependent variables. Since the Universe is homogeneous, and geometry is determined by the presence of physical objects, the geometry of every region of sufficient size (viz. a region in which there are enough clusters for their random motions and distribution to cancel out) will be the same as that of any other such region. This is measured by k. If $k = 0$ or -1 we can extrapolate without limit; beyond every region of finite size there will be another one and so the Universe will be infinite. If $k = +1$, there will be only a finite number of regions and the geometry will be elliptical (or spherical). The volume of an elliptic universe is $\pi^2 R(t)^3$. (Of a spherical universe it is $2\pi^2 R(t)^3$.)

Mathematical cosmologists have produced from the simplified field equations various more specific 'models', by making assumptions about the values of the constants and for the present cosmic instant of the variables referred to above, thereby presenting specific cosmological theories ready for confirmation or rejection by the results of astronomical observation. Thus if we suppose, as Einstein reasonably did in 1917 before the evidence from the red-shift of the expansion of the Universe had been produced, a static Universe, $R'(t) = R''(t) = 0$. The equations above then reduced to:

$$\frac{8\pi G}{c^4} p = \lambda - \frac{k}{R(t)^2}$$

$$-\frac{8\pi G}{c^2} \rho = \lambda - \frac{3k}{R(t)^2}$$

Eliminating λ, we obtain

$$R(t)^2 = \frac{kc^4}{4\pi G(p + \rho c^2)}$$

Since $p \geqslant 0$ and $\rho > 0$, $k = +1$. $R(t)$ has a value determinable by the values of pressure and density from the above equation. Hence if the Universe were static and if Einstein's field equations in the simplified form given on p. 239 were correct, Space would be elliptical (or spherical).

Different solutions can be obtained by making different assumptions about the values of the constants and variables. If we take the red-shift to indicate current expansion of the Universe, $R'(t)$ must be positive for the present cosmic instant. Thus if we suppose, as suggested by observation on galactic recession, that $R'(t) = \frac{R(t)}{T}$, where T is a constant, we get the De Sitter model with $p = \rho = k = 0$. If we propose $p = \lambda = k = 0$ we get the Einstein–De Sitter model where $R(t) = at^{\frac{2}{3}}$, a being a constant ($p = 0$ and $\rho = 0$ should be interpreted as claiming that pressure and density respectively are insignificantly small). Many of these models, it should be noted, are quite compatible with other general theories of mechanics than the General Theory of Relativity.

Although, as I have written, there is quite a lot of experimental evidence in favour of General Relativity, there are these rival general theories of physics, which are by no means ruled out. Most of these are slight variants on General Relativity. A few are different in structure. Thus in 1952 J. L. Synge extended Whitehead's theory of gravitation in a way that gave predictions of the advance of Mercury's perihelion and the deflection of light-rays near the Sun differing only insignificantly from Einstein's. There are also theories of greater generality than General Relativity, such as the theory of supergravity of D. Z. Freedman, P. van Nieuwenhuizen and others, which seek to reduce the four forces known to physics — gravity, electromagnetism, the 'weak' force, and the 'strong' nuclear force — to one, thereby providing a unified account of all physical phenomena. This theory makes largely the same predictions as

General Relativity within the range to which both are applicable, but it makes some different predictions.[1]

Alternative theories of mechanics when combined with the Robertson–Walker line element will yield different cosmological theories. The most important of these in recent years has been the Bondi–Gold–Hoyle Steady State Theory. These writers set up, as a postulate, the perfect cosmological principle and understand by it[2] that in any sufficiently large region of Space at all cosmic instants the density, energy, and distribution patterns of galactic clusters are the same. The continual expansion of the universe is compensated by the continual 'creation' of new matter. To give precise mathematical form to this theory, Hoyle modified the equations of General Relativity by abandoning Einstein's λ term and introducing instead a new 'creation' tensor $C_{\mu\nu}$ into the equations. This yields

$$R_{\mu\nu} - \tfrac{1}{2} R g_{\mu\nu} + C_{\mu\nu} = -8\pi G\, T_{\mu\nu}$$

Given the rate of expansion of the Universe suggested by observation on galactic recession $R'(t) = \dfrac{R(t)}{T}$, where T is a constant, this yields $R(t) = R_0 e^{t/T}$. We assume, as the simplest supposition, Euclidean geometry $k = 0$. We deduce a rate of 'creation' of matter $Q = \dfrac{3\rho}{T}$. Taking the best ascertainable value of T this yields a rate of creation of matter $Q = 10^{-15 \pm 2}$ hydrogen atoms/cm^3 per year.

Further, a cosmological theory may be set up inconsistent with the Robertson–Walker line element. If the evidence adduced in Chapter 11 for the applicability of the Robertson–Walker line element, including the evidence for the homogeneity of the Universe, still stood, the new theory to be acceptable would have to explain that evidence and also account for other astronomical data more adequately than could theories presupposing the Robertson–Walker metric.

In these ways a variety of cosmological theories can be and have been set up. In so far as a theory presupposes a system of mechanics and optics which makes predictions about local phenomena, viz. phenomena

---

[1] For an elementary account of this work and references to more technical accounts, see D. Z. Freedman and P. van Nieuwenhuizen 'Supergravity and the Unification of the Laws of Physics', *Scientific American*, February 1978, 126–143.

[2] These and other writers sometimes, as we saw in note 1 to p. 195, take the principle in a more general sense.

occurring near the surface of the Earth, it must account satisfactorily for those phenomena. Likewise astronomical phenomena must be as predicted by the theory, if it is to be accepted. Among theories compatible with observed phenomena, the best supported and the one provisionally to be adopted is the simplest and most coherent one.

A difficulty however arises on the astronomical scale that in order to ascertain what phenomena in fact occur we often need the cosmological theory which has not yet been established. Cosmological theories, as we have seen, make predictions about the density of matter (ρ) and rate of mutual recession of clusters ($R'(t)$ at cosmic instant t) and other apparently observable features of the Universe. But the difficulty is that in order to establish of any region of Space at any instant what the values of these are, we have often to assume a cosmological theory. (On the local scale this difficulty does not arise nearly so readily.)

Thus cosmological theories make predictions about the value of ρ, the density of matter, in all regions. Yet in order to ascertain how close together are distant bodies which appear tangentially close, we have to know whether light travels in straight lines and what is the curvature of Space. Locally light travels in straight lines, and we have no reason for supposing that its path is other in more distant regions. Yet cosmological theory leaves it open what is the curvature of Space on the cosmic scale. If two bodies at a small distance d from Earth have a certain angular separation and two distant bodies at a distance D from Earth have the same angular separation, then the distance of the distant bodies from each other will be $\dfrac{D}{d}$ times the distance of the smaller bodies from each other if the geometry is Euclidean, but greater than that amount if the geometry is of negative curvature. Hence we need to know the curvature of Space before we can calculate from observations the density of matter in distant regions. Again, we observe distant bodies as they were some period of time ago. Their density may differ for this reason from that of nearer bodies. In order to test our cosmological theory, we have to know how long past are the states in which we are observing them. But to know that we have to know their distance from us and the velocity of light. We have already seen in Chapter 5 how estimates of distance presuppose cosmological theory; and in Chapter 11 how the formula for the velocity of light depends on $R'(t)$.

What applies of ρ applies to the other terms which seem equally directly related to observation. Thus $R'(t)$ indicates the rate of mutual recession of the clusters at cosmic instant t; yet we can only calculate this from observations of the red-shift if we know how far away and how long ago are the clusters studied. Yet to calculate this we need the cosmological theory which we are seeking to establish.

Theory and observation thus interact very closely in cosmology. This does not mean that the whole business of theory construction is circular but it does mean that it is rather indirect. We have already schematised in Chapter 4 the process of interaction between theory and observation in establishing the geometry of space, using only primary tests. The use of secondary tests makes the process yet more circuitous, but the principles are the same. The business of theory construction in cosmology can be schematised as follows. We interpret our observations by making assumptions about the values of such terms as k in order to obtain data about phenomena such as the density of matter or rate of recession of the galactic clusters in various regions. We then make different alternative sets of assumptions and interpret our observations with their aid. We then construct the simplest cosmological theory compatible with each set of data obtained from observations interpreted under certain assumptions and compatible with those assumptions. Thus if we make our estimates of the distance of clusters by assuming that $k = 0$, any theory postulated to explain the data calculated in this way must claim that $k = 0$. From the rival simplest theories we then choose the theory which is overall the simplest. That theory is the one which is most probably true.

These considerations should make it clear why it would not be very easy to show that, despite the evidence adduced in Chapter 11, the Universe was not homogeneous at all instants by the clocks of each cluster including our own, or to produce further support from observations for this claim. It might appear that we have only to study distant regions of Space and see if there are as many clusters per unit volume as in a local region. Suppose we scan the sky with a telescope and find that there are very few very faint clusters observable. This certainly suggests that there are very few very distant clusters and hence that the density of matter is smaller in distant regions than in our own. But there are other ways of interpreting the observations. Perhaps the more distant the clusters, the more they are obscured by intergalactic dust. The clusters are observed as they were a long interval of time ago. Perhaps clusters were not then so clearly defined; there was the same amount of matter as now but it was more spread out and so would be less easy to detect in a telescope. Perhaps Space has positive curvature and so clusters are not as far apart as they seem, and so on and so on. However, the various alternative ways of interpreting the observations and the data obtained in these ways might give rise to no theories simpler than or as simple as the theory that the Universe was not always homogeneous. In that case that theory would be the theory best supported by observations. So astronomical observations other than those cited in Chapter 11 could provide evidence for or against the homogeneity of the Universe at all instants by the clocks of each cluster, and so the applicability to it of the Robertson–Walker line element.

We noted in Chapter 9 that extrapolation from known spatio-temporal regions to unknown became less justifiable the more distant in space and time were the unknown regions from the known. If I have observed many regions within a thousand million parsecs of the Earth, I can speak with confidence of the properties of regions between 1000 and 2000 million parsecs from the Earth, but with less and less confidence of increasingly distant regions. In regions billions of millions of parsecs away there may be no clusters, or clusters not receding from us but approaching us, and so if only our telescopes were good enough they would detect their violet-shift. If, however, extrapolation from known regions of space to near-by unknown regions leads us to conclude that these are all the regions that there are, viz. that the space is closed and so the Universe s- (and so m-) finite, inference from the properties of some regions of the Universe to the properties of the Universe as a whole would seem a well-justified procedure. But extrapolation from the properties of a finite number of regions to the properties of an infinite number of regions of equal volume is clearly a more shaky process. Yet we must make this extrapolation if we are to conclude on the basis of the properties of an observed region (e.g. from the fact that the geometry of that region is Euclidean or hyperbolic) that there is no limit to the number of regions of similar volume and so that the Universe is open and infinite.

Nevertheless, however shaky the inferences may be, clearly my inferences, based on the assumption that in unobserved regions traces will function as in observed regions, about any region however distant will be more justified than any random guess of equal detail which I might make about the character of those regions. In the Sahara desert case considered in Chapter 9, if I conclude that 2000 miles away in a certain direction there is a square mile of sandy desert my claim, however weakly justified, will be more justified on the evidence available to me than if I claim that there is there a mountain of green cheese or an expanse of water. So it is with the cosmology case. However distant the regions of space and time about which I make inference by traces, my inference to their character will be more justified than any random guess of equal detail which I might make about that character. It is characteristic of science to infer the properties of all members of a possibly infinite class from those of a small finite class — of all pieces of copper from those of a small finite number of pieces of copper.[1] I conclude that justified inference to the spatial finitude of the Universe and to the other properties of the Universe as a whole if the evidence is that it is *s*-finite, would be justified much more strongly than justified inference to the *s*-infinitude of the Universe and to the other properties of the Universe as a whole if the evidence is that it is *s*-infinite.

So far astronomical investigations have yielded insufficient well-established data for a cosmological theory to be well established against rivals. Some of the cosmological 'models' which have been developed are clearly ruled out by astronomical investigation. Thus Einstein's original model of a static Universe and Steady State Theory in its original form have both been rejected on the basis of astronomical evidence. The redshift indicated that the Universe was not static, and investigations into the density of matter in distant regions gave results inconsistent with Steady State Theory. To choose between other cosmological theories we wait for more astronomical and local physical observations and we seek compatible with them the most simple overall theory of the distribution and interaction of the constituents of the Universe, giving good predictions. It will be a long time, if ever, before such a theory clearly emerges way ahead of rivals. Meanwhile among almost equally well substantiated theories are some in which $k = +1$, some in which $k = 0$, and some in which $k = -1$. Hence we must for the moment remain agnostic about the size and geometry of the Universe.

BIBLIOGRAPHY

[1] I. Kant, *Critique of Pure Reason*, Transcendental Dialectic, ch. 2. 'The Antimony of Pure Reason', especially B.454–61.

For Einstein's original papers on General Relativity and Cosmology, see:
[2] A. Einstein, 'The Foundation of the General Theory of Relativity' and 'Cosmological Considerations on the General Theory of Relativity' reprinted in A. Einstein *et al.*, *The Principle of Relativity* (trans. W. Perrett and G. B. Jeffery) London, 1923.

[1] Contrary to the apparent practice of most scientists, some philosophers of science claim that no finite collection of data could substantiate to any degree a hypothesis about a possibly infinite class. See R. Carnap, *Logical Foundations of Probability*, 2nd end., Chicago, 1962, 570–5, and M. Hesse, *The Structure of Scientific Inference*, London, 1974, ch. 8.

For simple descriptions of the structure of modern Cosmology, see:

[3] M. Berry, *Principles of Cosmology and Gravitation*, Cambridge, 1976.

[4] G. C. McVittie, *Fact and Theory in Cosmology*, London, 1961, especially chapters 4–7.

For modern criticism of the legitimacy of all talk about the properties of the Universe as a whole, see references [6], [7], and [8] to Chapter 15. I claim to have met in the text all the points made by these writers, though I have not discussed them by name individually.

15 The Beginning and End of the Universe

What does it mean to say that the Universe had a beginning or will have an end and of what sort of proof are these propositions susceptible?

We understand, as in the last chapter, by 'the Universe', all physical objects spatially related to the Earth taken together. To say that the Universe had a beginning is to say that a finite period of time ago there were no such objects; and to say that the Universe will have an end is to say that a finite period of time ahead there will be no such objects. To say that the Universe had a beginning is, if there is Absolute Time, viz. a cosmic time scale, to say that on the cosmic time scale there were no physical objects spatially related to the Earth at an instant n units of that time scale ago, where n is a finite number. If there is no cosmic time scale, but a number of different time scales resulting from rival accounts of the laws of nature equally simple and successful in predicting actual and possible observations, then to say that the Universe had a beginning is to say that there were no such objects at an instant a finite number of units of time ago on each of these admissible time scales. To say that the Universe had no beginning, viz. has existed eternally, is, if there is a cosmic time scale, to say that no finite number of units ago on that time scale were there no such objects. If science were found to use more than one time scale, to say that the Universe has existed eternally is to say that on no admissible time scale was there an instant, a finite time ago, at which there were no physical objects spatially related to the Earth. If on one admissible time scale, there were no such objects a finite time ago, whereas on some other admissible scale there was no instant, a finite time ago, at which there were no such objects, then assertions that the Universe has or does not have a beginning fail of significance. Such a situation arose in the cosmology constructed by E. A. Milne, referred to on p. 180. Milne's t-scale, taken by itself, yielded a finite age to the Universe, whereas the τ-scale yielded an infinite age. Milne himself did not continue to advocate the applicability of the two scales, and fortunately science since his time has proceeded, as we have seen in Chapter 11, in the belief in the validity of a cosmic time scale.

This analysis of the meaning of the statement that the Universe had a beginning and of the statement that the Universe has lasted eternally applies, *mutatis mutandis*, to the meaning of the statement that the Universe will have an end and of the statement that the Universe will last for ever.

Let us now consider what sort of proof could be given for the propositions that the Universe had a beginning or, alternatively, has lasted eternally. As with the parallel case of the spatial finitude or infinitude of the Universe, there have been in the history of philosophy a number of arguments purporting to show that of logical necessity a certain conclusion about this topic is true. Such arguments, as in the last chapter, I shall term *a priori* arguments. The *a priori* type of argument to show that the Universe is eternal is adumbrated in Aristotle [1]. *A priori* arguments in rebuttal of Aristotle, to show that the Universe had a beginning a finite time ago, are found in St Bonaventure [2]. Christian theology has of course always maintained that the Universe did have a beginning at a time. But most Christian theologians before St Bonaventure appealed to revelation in justification, though they did not in general find it necessary to justify what was until the thirteenth century a fairly uncontroversial belief. (I shall not discuss the validity of arguments from revelation, since clearly too many special considerations are involved here.)

St Thomas Aquinas [3] argued against Aristotle and St Bonaventure that all arguments known to him in favour of either position were invalid. Kant subsequently claimed, for reasons similar to those discussed in the last chapter for the corresponding spatial claims, not merely that all the arguments known to him in favour of either position were invalid, but that any argument to prove that the Universe did or did not have a beginning must be invalid, for any talk of the Universe as a whole fails of significance. As with the corresponding spatial case, the *a posteriori* type of argument, though not unknown in medieval philosophy, only becomes of importance from the nineteenth century onward. *A posteriori* arguments of any plausibility today rely on the achievements of modern scientific cosmology. They point out certain observable features of the Universe which might not have held and then argue that these can only be accounted for if we suppose that the Universe had a beginning, or, alternatively, if we suppose that the Universe has lasted eternally.

I shall now consider the two most common *a priori* arguments in favour of the two positions. Like Aquinas and Kant, I shall urge that both are invalid.

The best-known *a priori* argument that the Universe had a beginning a finite time ago is the one which Kant gives in favour of the thesis of this First Antinomy. This is as follows.

> If we assume that the Universe has no beginning in time; then up to any given moment an eternity has elapsed, and there has passed away in the world an infinite series of successive states of things. Now the infinity of a series consists in the fact that it can never be completed through any successive synthesis. It thus follows that it is impossible for an infinite world-series to have passed away, and that a beginning of the world is therefore a necessary condition of the world's existence ([4], B.454).

Kant claims that, but for the inadmissibility of talking about the Universe as a whole, this argument would be valid. This, however, seems plainly false, and so there is no justification for taking Kant's escape route. The argument fails for the following reason. In considering series we normally suppose the first member identified and then go on to see what can be said about the other members or the sum. Hence it seems reasonable enough to claim that a completed infinity is impossible. But if it makes sense (which seems dubious) to talk about a series with a last, but not a first, member identified, the impossibility of a completed infinite series of this type is in no way obvious. It seems more natural, however, to say that the series to be considered in discussing our question is a series of causes, not effects, and that the first member is the present state of the Universe. In this case there is no difficulty in supposing that the series may be infinite, for it may have no last member and so it would not be completed.

The best-known *a priori* argument for the eternity of the Universe derives from Aristotle.[1] The medieval theologians discussed it and claimed to have refuted it, and Kant did the same. The argument is that if the Universe had a beginning at an instant of time, before that instant there must have been nothing at all. But then nothing cannot give birth to something. The creation of the Universe needs to be explained by some preceding state which brought it about. But if all there was before creation was void, then there was no state to explain the creation. Therefore creation cannot have occurred. This is roughly the argument

---

[1] [1] 280a. Less satisfactory arguments occur later in [1].

that every event or state needs a preceding event or state as cause to explain its occurrence or continuation. But an initial state of the Universe cannot have such a cause (if we ignore, in view of the lack of evidence for its existence apart from evidence which brings in theological considerations, the possibility of another space, an event in which could be the cause of the existence of our Universe). Therefore, there cannot be such an initial state. This argument is given by Kant in his discussion of the First Antinomy as the argument of the antithesis, if we take his argument in one of the two possible ways of taking it.[1] Kant's only objection to the argument of his antithesis is that considered in the last chapter, that we cannot talk of the Universe as a whole, mainly because if we could the thesis would also be valid.

Again there seems no need to take such a radical line to rebut the argument. The argument assumes as a logically necessary truth that every state has a cause in the sense of a preceding state which brings it about. (This proposition is normally phrased 'every event has a cause' but as its proponents would wish to see it as claiming that not merely every change of state of physical objects but that every continuation of state of physical objects needed a cause, I shall phrase it in future as I have done here.) Kant claims that this is in his sense an *a priori* truth and his argument only works as stated if this claim is right. I wish to argue on the contrary that 'every state has a cause' is a logically contingent proposition.

A supporter of the proposition can point to evidence in favour of it. The vast success of science in explaining the hitherto inexplicable suggests that one day science will explain everything. But since explanation consists in finding causes, if we are to provide explanations the causes must be there to be found. The success of science shows that these are there. An opponent of the proposition can point to evidence against it — the success of the Quantum Theory. If the Quantum Theory be true, as experiments seem to suggest, then there will be a vast range of phenomena for which precise explanations cannot ever be given. If no cause of some state of physical objects could ever be found, then is not that the reason (whether or not conclusive reason) for supposing that the state had no cause? An advocate of the logical necessity of the truth of

---

[1] [4], B.455. Alternatively we may understand Kant as arguing that the Universe could not have begun at an instant because that instant would have been undatable by reference to an external time scale, viz. preceding events. But it would have been datable by an internal time scale, viz. by the interval of time ago from the present instant. See [5] for discussion of this form of the argument, and Kant's treatment of the First Antinomy in general.

the proposition 'every state has a cause' has to say that all these reasons are entirely irrelevant to the truth of the proposition; that a man would have just as much reason for believing it to be true in an apparently vague chaotic Universe for which no precise laws had ever been formulated as in our Universe before the first work on Quantum Theory had been done. That position seems implausible.

I conclude that the *a priori* argument for the eternity of the Universe needs further substantiation, since it cannot be taken as a logically necessary truth that every state has a cause.[1] Of course even if the argument were valid, it would only support Kant's doctrine that all talk about the Universe as a whole is improper if the argument of Kant's thesis was also valid, and we have concluded that it is not. If the argument of the antithesis taken as we have taken it or in the alternative way alone were valid, it would show what it purports to show, that the Universe has lasted eternally. But, as it stands, it is not valid.

Now certainly in view of the vast success of science to date, the scientist in his search for causes is justified in hoping that one can be found for any given state, unless he has good grounds in a particular case — as with phenomena within the quantum range — for doubting this. The cosmologist is justified in making an initial working assumption that every state of the Universe can be explained in terms of a preceding state. But he might find grounds for believing that such an explanation could not be given. Some of the states of the Universe might be ones which could not have come into existence by any known scientific process, and the postulation of a new process might produce chaos in scientific theory. Since the scientist always takes as the true theory a very simple rather than a very complicated theory, when both could serve to explain almost all known states, he would not adopt the very complicated theory merely in order to explain one or two otherwise inexplicable states. Hence it cannot be taken for granted that there are no states of the Universe inexplicable in terms of preceding causes.

---

[1] Aquinas argued that although it was a necessary truth that every state had a cause, that cause need not be another state of physical objects. He allowed that instead an agent might be the cause ([3] 2.35–7). Science, however, operates with state or event causality and attempts to explain all states in terms of other states which bring them about. The suggestion that agent causality is a different type of causality from state or event causality, but that explanation in terms of it is equally good explanation, is a difficult and much disputed one. To discuss, however, would take us too far afield. We have accepted Aquinas's claim that is it not a logically necessary truth that every state has a precedent state as a cause; which alone is the relevant point here. For the argument of Kant's antithesis depends on the latter being a logically necessary truth.

Alternatively the cosmologist might succeed in giving an explanation of every state of the Universe in terms of a preceding state, and that would show that that sort of explanation could be given.

All this means that the claim that every state of the Universe is caused by a preceding state needs substantiating by detailed scientific cosmology. So we must turn from the *a priori* arguments to the *a posteriori* arguments about the beginning of the Universe, these being ones which include among their premisses statements of particular observable features of the Universe. We must look, in other words, at models of the Universe developed from the General Theory of Relativity or other general theories of mechanics. We investigated in the last chapter the evidence available now and the type of evidence which we could have subsequently in support of some one of such models. Before examining cosmological models, however, we had better clarify the structure of the *a posteriori* arguments used in both cases, to prove that the Universe had a beginning and to prove its eternity.

The *a posteriori* arguments to prove that the Universe had a beginning are developed by popularisers of science or theology from what I may term creation theories of science. All creation theories have the following common structure. The Universe is said to be now in a state S_p. It is further claimed that a fundamental law of nature controls the evolution of its states, such that if it is now in state S_p, then so many years ago it must have been in state S_m and in so many years' time it will be in state S_n. S_p is thus a trace of those states. On the basis of this law, inference is made that at t_0, some finite number of years ago on all admissible time scales, viz. normally on a cosmic time scale, the Universe must have been in a state S_0. But S_0 is either a state known, on the basis of this or other laws of nature, to be physically impossible or is a state in which there is nothing physical existent. But if the state is physically impossible or there is nothing physical existent, the popular argument takes over from the scientific, then the Universe cannot have existed at that instant, but must have come into existence subsequently.

Scientific laws, we saw in Chapter 9, are forward moving; they state the future consequences of some present state. Hence if a law is to be used for retrodiction we have to have reason to believe that any state which could have caused the present state other than the one stated by the law did not occur at the relevant time. However, in the case of a purported cosmological law, that is, a purported basic law governing the succession of states of the Universe, given that there can be no cause lying outside the Universe of a state of the Universe, then unless the purported law is not completely true or there is an uncaused state of the

Universe, it can be used for retrodiction. For the Universe is all the physical objects that there are spatially related to the Earth. If there is a cause of a state of the Universe which lies outside it, it could only be a state of an object not spatially related to the Earth. But we would hardly be justified in postulating the existence of such objects solely to avoid the conclusion that an uncaused state occurred. If we suppose as a working hypothesis — to avoid arguments about theology — that we have no evidence of the existence of such states, we must look for causes of states of the Universe within the Universe. But in that case if some state of the Universe S_m has a cause, it must be a precedent state of the Universe, say S_k. But then either the purported cosmological law in question permits S_k to have brought about S_m or it does not. If it does not it cannot be the true law governing the evolution of states of the Universe. If we have evidence that a state of the Universe is brought about other than in the way which we could retrodict by using a suggested cosmological law, then it shows that the cosmological law must be wrong. The case with cosmological laws is thus radically different from the case with other laws. Men walking on sand produce footprints. We see certain footprints and infer that men have walked. But suppose we have evidence that these footprints have other causes (say, animals walking with man's shoes on them). That has no tendency to show that the purported law 'men walking on sand produce footprints' is false. It is otherwise with cosmological laws (given the non-existence of other spaces), for states of the Universe can only have other states of the Universe as causes. If the purported cosmological law compels us to retrodict S_j from S_m, but we have evidence that S_k, not S_j brought about S_m, that shows that the purported law is wrong. The only way in which we can avoid the retrodictive consequences of accepting a cosmological law is by postulating that some state of the Universe had no preceding state as cause.

The prototype of the *a posteriori* argument to a beginning of the Universe is the thermodynamic argument used by various popularisers of science and theology during the last hundred years. The argument runs as follows. The Universe is at present in a state of considerable thermodynamic disequilibrium, and entropy lies between its minimum and maximum. By the second law of thermodynamics, as originally formulated, entropy increases or remains constant, never decreases in a closed system. The Universe is a closed system since it consists of all the physical objects that there are spatially related to the Earth. Therefore some finite number of years ago, n years, entropy must have been at its minimum. But it could not have been like this for ever previously, since a long continuing state of minimum entropy (S_0) is a physically impossible state. Hence there could not have been a Universe more than n — or perhaps $n + 1$ — years ago. Now this argument is not much in vogue today because it begs many important scientific questions — for example, it may only work for a spatially finite Universe, for the existence of which it produces no evidence. Further, on the statistical version of the second law (see Chapter 13) decreasing entropy is not completely ruled out and it is as probable that an eternally closed system should be at any instant in a state of decreasing entropy as that it should be in a state of increasing entropy. But I cite the argument because it illustrates clearly and simply the *a posteriori* arguments for a finite age of the Universe which are in vogue today.

The *a posteriori* arguments in vogue today are based on the cosmological 'models', viz. theories of the Universe substantiated by various empirical observations in the way described in Chapter 14. They are, as we noted, often constructed by giving various values to the constants and variables of equations developed from the General Theory of Relativity or rival theories of mechanics. The resulting solutions yield various values of $R(t)$, the scale factor such that if the distance between two galactic clusters at cosmic instant t_1 is $R(t_1)x$, then the distance between them at t_2 will be $R(t_2)x$. $R'(t)$ is the rate of expansion of the Universe. We get oscillating Universes, and universes with constant or variable rates of expansion. Some of the 'models' yield an instant t_0, a finite time ago at which $R(t_0) = 0$. They show, that is, expansion from an infinitely dense Universe. But an infinitely dense Universe is a physically impossible Universe. Hence the Universe must have come into existence after t_0. The argument here is more subtle than the thermodynamic argument, but its structure is the same. The Universe is now in a certain state — S_p, that is, in this case, has a certain scale. The law stating its expansion is a fundamental law of physics. Hence a first state of the Universe must have been subsequent to t_0. This is the conclusion to which we are led if a model yielding an instant t_0 at which $R(t_0) = 0$ is very well substantiated. For if the state of the Universe at the instant immediately after t_0, S_1 at t_1, was brought about by a causal process, it cannot have been in accordance with the fundamental theory stated. If an explanation of S_0 is to be provided, it can only be done by having a different cosmological theory. The new theory would need justification, more justification than that the old theory could not explain S_0. For if the latter sort of fact were always sufficient justification for accepting a new theory, then 'every state has a cause' must be a necessary truth — and I have argued that it is not. If the mere failure of a theory to provide a cause for something within its domain shows that the theory is mistaken then the false consequence follows.

Hence we can only maintain the existence of the Universe before t_0 by a supposition of entirely different fundamental laws of the Universe for which there is — *ex hypothesi* — no other evidence from its states. The criterion of simplicity demands simple laws with one unexplained state rather than complex laws explaining all states. Scientific evidence in consequence points to S_1 being a state uncaused by a precedent state. If S_1 was not caused, it could not have been preceded. For given the truth of the current cosmological theory, any preceding state would have brought about some other state than S_1. But S_1 occurred. Therefore it cannot have been preceded. So the Universe must have had a beginning a finite time ago.

It is not mere failure to achieve an explanation of S_1, that provides evidence that S_1 could not be explained. Science has often failed to explain something for centuries and then succeeded. But the evidence that S_1 could not be explained would be that such explanation was ruled out if you accepted the very successful theory which so brilliantly accounted for all other cosmological phenomena (viz. all the other states of the Universe). This point is brought out by the similar case of Quantum Theory. It is not the mere fact that scientists have not yet succeeded in giving a complete explanation of certain phenomena within the quantum range which makes them want to say that those phenomena are not fully explicable. It is the fact that they have succeeded in explaining very well to a certain degree of accuracy a whole range of atomic phenomena, and that the theory by which they explain those phenomena rules out any complete explanation of phenomena within the quantum range.

Arguments to prove that the Universe has lasted eternally have the following common pattern. The Universe is said to be now in a state S_p. It is claimed that a fundamental law of nature controls the evolution of that state, such that if the Universe is now in state S_p, then so many years ago it must have been in state S_m and in so many years' time it will be in state S_n. On the basis of this law retrodiction is possible to physically possible states as long ago on all admissible time scales, viz. normally the cosmic time scale, as you choose. So S_p is best explained by postulating a previous state S_m, and S_m is best explained by postulating a prior state S_j and so on *ad infinitum*. If you accept that the evidence shows this, then you accept that the best explanation of the present state of the Universe is as a member of an infinite series of such states.

Various cosmological theories do permit infinite regress of explanation. The best known is Steady State Theory. As explained in Chapter 14, according to this theory the Universe has always consisted of clusters of galaxies at roughly the distance apart that they are now from each other. The increasing distance apart of any two clusters is compensated by the continual 'creation' of matter in the space between clusters. This new matter then condenses into new clusters, and hence average distance between clusters remains roughly constant. Other theories permitting infinite regress of explanation include oscillating Universe theories (that is, theories of the Universe oscillating between physically possible states, viz. ones both having a finite positive value of $R(t)$). According to oscillating universe theories the Universe is now expanding, was previously contracting and so on into the infinite past or future. On all such theories a present state of the Universe is best explained by a preceding state and so *ad infinitum*.

Now all *a posteriori* arguments to the eternity or to a beginning of the Universe, like *a posteriori* arguments to its spatial finitude or infinitude, depend on the truth of a cosmological theory which, as a scientific theory, may be challenged on the usual scientific grounds. It may be urged, first, that the Universe is not now in the state S_p — that the claim that it is represents an illegitimate interpretation of observations. We analysed in the last chapter the difficulties involved in interpreting astronomical observations. Secondly — as is most usually done — it may be claimed that the proposed law of change of S, the proposed cosmological theory, is wrongly stated; or, if correctly stated, is not a fundamental law of nature, but applies only over a certain temporal period. The latter is the claim that its application in certain conditions is a derivative consequence of more fundamental laws of nature; and since these conditions only hold for a certain period, the law in question is only operative for that period. All proposed cosmological theories are challenged on this ground. Thus G. C. McVittie claimed of models having $R = 0$ at an initial instant t that: 'Going backwards in time we can legitimately expect that, long before $R = 0$ is reached, conditions similar to those that prevail at present in the observed universe would have broken down. Thus the uniform model universes would cease to apply.'[1] The conditions under which the proposed cosmological theory applies are matter for normal scientific argument. What must be shown is that the theory is the simplest theory of the Universe consistent with observations so far made, and hence that there are grounds for

---

[1] G. C. McVittie, *Fact and Theory in Cosmology* (London, 1961) pp. 151f.

supposing that it is so fundamental that it would still hold under very different conditions of the Universe. Thirdly, the theory may be challenged on the grounds that the consequence that at an instant t_0 there is a state S_0 which is a state without anything physical existent or a physically impossible state does not follow from the premises. This would be a matter for mathematicians to clarify.

But given that the proposed cosmological theory is in fact the simplest theory consistent with observed phenomena, the consequences for the arguments for a beginning or for the eternity of the Universe do follow. However, for reasons given in Chapter 9, justified arguments to a beginning of the Universe will in one respect be much stronger than justified arguments to its eternity, just as, we found in Chapter 14, justified arguments to its spatial finitude are as such much stronger than justified arguments to its spatial infinitude. The point is that we have obtained our scientific theory by extrapolation from a small temporal period. We construct the simplest theory consistent with the observed data and extrapolate these backwards to other temporal periods. To establish our conclusion that the Universe had a beginning we only need to extrapolate backwards for a certain finite temporal period (e.g. about 14,000 million years, on most current relevant theories). But to establish the conclusion about the eternity of the Universe, we have to take the behaviour of objects in the region which we have observed as typical of the behaviour of objects at all temporal periods, however distant from our own. The present behaviour of clusters might be due to some peculiar feature of current cluster distribution brought about by some event in the quite remote past, and hence they might behave in a quite different way under different circumstances, e.g. in the yet more remote past, or again we might have misdescribed the present state of the Universe very slightly. The slightest error in interpreting our observations might not make any difference to our inference to the state of the Universe 14,000 million years ago, but a vast difference to our inference to its state trillions of trillions of years ago.

On the other hand, in order to retrodict that there were no physical objects spatially related to the Earth more than a certain number of years ago, we must claim knowledge of the present state of the whole Universe. Only if we can show that no physical objects now or previously spatially related to the Earth existed at a certain prior instant, can we conclude that the Universe had a beginning. It is not sufficient to show that in a certain region R there were no physical objects a certain period of time ago. Whereas to show that the Universe has existed eternally we have only to show of one region of it that there have always been physical objects in it. Hence a spatial extrapolation is necessary to the present state of the whole Universe before we can make our retrodiction to prove that the Universe had a beginning. If the spatial extrapolation indicated, in the way described in the last chapter, that the Universe is spatially infinite, any conclusion that it had a beginning would be as weakly justified as the conclusion that it had lasted eternally since both would depend on a process of infinite extrapolation. Only if the evidence suggested that the Universe was spatially finite would inference to its temporal beginning be well justified, for only then would the conclusion not depend on a process of infinite extrapolation. Otherwise either conclusion, if substantiated, would only be very weakly substantiated.

As we showed in the last chapter, cosmology is a long way yet from being able to provide us with a well-established cosmological theory to the exclusion of rivals. However, there is no doubt that the models best substantiated today are ones which show the Universe expanding from a 'big bang' some 14,000 million years ago. These models successfully predict not merely the density and rate of recession of the galaxies, but the ratios of the various chemical elements to each other and to radiation in the universe, and above all the background cosmic radiation. This is electro-magnetic radiation in the microwave region of the spectrum (i.e. with waves much longer than light waves) streaming from all directions. It is most plausibly explained as a cooled red-shifted remnant of the primeval fireball. But whether such an explosion can be explained as a result of a previous contraction of the Universe is unclear. We have however been investigating in advance the metaphysical consequences of supposing that it cannot be thus explained, and that by far the best substantiated theory will be one which does not allow infinite regress of explanation.

What next of the end of the Universe? Could one show that the Universe will have an end or alternatively that it will exist eternally? *A priori* arguments on this question hardly exist — philosophers have generally ignored it. There is no plausible parallel to the *a priori* argument that the Universe must have had a beginning, but there is an initially plausible parallel to the argument that the Universe must have lasted eternally. This would be an argument that since every state must have an effect, a state produced by the precedent one, every state of the Universe must be succeeded by another. But the proposition that every state must have an effect is, like the proposition that every state must have a cause, not a necessary truth — and for the same reasons. It is conceivable that only some states of objects bring about other states.

That every state brings about other states is something that needs substantiating by showing that the simplest scientific theories compatible with phenomena predict infinite successions of states.

So let us consider if *a posteriori* arguments could show grounds for believing that the Universe will have an end or alternatively will last for ever. The proposition that the Universe will last for ever could certainly be substantiated, albeit very weakly, by *a posteriori* arguments. If a cosmological theory can be substantiated which shows that, starting from the present state S_1, all subsequent states on all admissible time scales, viz. normally on a cosmic time scale, S_2, S_3, etc. were physically possible states of physical objects and that there was no limit to prediction of such states by the theory, then the theory shows that the Universe will go on for ever. All currently viable cosmological theories have this characteristic (apart from some — see note 1 to p. 260 — which only purport to hold over a finite temporal period). For example, the Einstein–De Sitter model which has the form $R(t) = at^{\frac{2}{3}}$ while admitting a beginning ($R = 0$ at $t = 0$) predicts an infinite future. In so far as such a theory is well substantiated, it gives grounds for affirming the future eternity of the Universe.

Such predictions to an infinite future suffer, however, from the weakness earlier discussed of all extrapolations from a finite temporal period to an infinite number of such. They also suffer from a further weakness of their own arising from the past/future sign asymmetry discussed in Chapter 13.

In order to predict an infinite future for the Universe, as to retrodict an infinite past, we need only make the inference for one region R. For if in R there always will be physical objects, then there always will be a Universe. In the case of retrodiction, this meant that, given the validity of the temporal extrapolation, no spatial extrapolation was necessary. However, in order to make a temporal extrapolation to the future of R we have to make a spatial extrapolation first. For we have reason to believe, as shown in Chapter 13, that while states of a system are reliable signs of past states of that system, they are only reliable signs of future states of a sub-system within that system. Hence only if we can claim knowledge of the state of a region wider than R at the present instant, can we predict the future state of R. Hence in order to predict the infinite future of a region of the Universe, we must claim knowledge of the present state of the whole Universe. If the evidence indicates that the latter is spatially infinite, our grounds for claiming knowledge of it will be, as we saw in Chapter 14, definitely weak. Hence to predict the infinite future of a region of an infinite Universe we should have to rely on two processes of extremely shaky inference. The inference will hence be doubly weak. If, however, we can infer that the Universe is spatially finite, the prediction to an infinite future of a region will be less weak, and so too therefore will be justified inference to the future eternity of the Universe.

The proposition that the Universe will have an end could only be substantiated by a peculiar form of cosmological theory. I earlier argued that the proposition that the Universe had a beginning could be substantiated either by showing that on the well established cosmological theory a finite period of time ago either the state of the Universe was a physically impossible one or there was nothing physical existent. Only the parallel to the latter form of theory could substantiate the doctrine that the Universe will have an end. For if the theory were such that it predicted a future physically impossible state of the Universe, that would only show that the theory was mistaken, not that the Universe would have an end. The difference between the case of the end and of the beginning of the Universe in this respect can be seen by recalling the form of scientific laws.[1] Scientific laws state the consequences of some state however it was produced. Hence if the proposed law predicts from the present state of the Universe a future physically impossible one, then, since such a state cannot occur, the law must be mistaken. Since scientific laws can only be used for retrodiction as we saw in Chapter 9, with further premises, a similar consequence does not arise in retrodictive inference. It is quite consistent to claim that a scientific law is true, from which can be deduced that if S_j occurred, S_m must have occurred, and to claim that S_m occurred but that S_j did not occur. But it is not consistent to claim that a scientific law is true from which it can be deduced that if S_j occurred, S_m must have occurred, and to claim that S_j occurred, but that S_m did not occur.

The only valid *a posteriori* argument to the end of the Universe would be one based on a well established cosmological theory which predicted a finite period of time ahead on all admissible time scales, viz. normally a cosmic time scale, a future state where there was nothing physical

---

[1] There are a number of models considered by cosmologists which have a future point singularity $R = 0$. These — for the reasons given above — cannot be taken as literally true scientific theories. Nor does any scientist so take them. They are regarded as inaccurate near the regions $R = 0$ and only used because they give accurate descriptions of present observable states and reasonable predictions and retrodictions of some near future and past states. They are mainly models of a Universe oscillating between values of $R = 0$ and finite values of R.

existent. I earlier referred briefly to the possibility of a theory from which it could be retrodicted that a finite time ago there was nothing physical existent. Using M for the mass of all the physical objects in any region of the Universe of a certain volume, the theory would be of the form $\frac{dM}{dt} = k$, where k was some positive constant or variable function of time. It could then be retrodicted (given a suitable form of k) that at some instant a finite period of time ago t_0, for all regions $M = 0$. Now if k was a suitable negative constant or function of time, the theory could predict a future instant when there would be nothing physical existent. Such a theory would therefore predict an end of the Universe. If it were well substantiated, we would have good reason to believe that the Universe would come to an end. To establish such a theory only a finite degree of temporal extrapolation would be necessary, although, if the evidence indicated a spatially infinite universe, an infinite spatial extrapolation would be needed.

Now in fact no such theory has, as far as I know, ever been put forward seriously by a physicist, let alone been well substantiated. The only recent theory which does not hold to the ancient physical principle of the conservation of matter (with energy) is the Steady State Theory, and that predicts the increase of matter. Since all current cosmological theories which are at all well supported by observation and experiment (apart from the few referred to in note 1 to p. 260 which only purport to hold over a finite temporal period) predict an infinite future for the Universe, I conclude that the conclusion of modern cosmology is that the Universe will go on for ever. However, in view of the need, described above, for at least one infinite extrapolation before this conclusion can be reached, we must add that evidence in support of it is weak.

BIBLIOGRAPHY

[1] Aristotle, *On the Heavens*, Book i, ch. 10; Book ii, ch. 1.
[2] St Bonaventure, *Commentary on the Sentences*, Book ii, dist. i, pars i, articulus i, quaestio ii.
[3] St Thomas Aquinas, *Summa Contra Gentiles*, 2.32–8.
[4] I. Kant, *The Critique of Pure Reason*, Transcendental Dialectic, ch. 2, 'The Antinomy of Pure Reason' especially B.454–61.

For discussion of Kant's argument, see:
[5] P. F. Strawson, *The Bounds of Sense*, London, 1966, part iii, ch. 3.

For description of the world-models of modern cosmology, see references in the Bibliography to Chapter 14.

For modern philosophical criticism of the coherence of talk about the Universe as a whole and of the possibility of reaching a conclusion about whether or not it had a beginning, see, among many writers who have a similar approach:

[6] M. Scriven, 'The Age of the Universe', *British Journal for the Philosophy of Science*, 1954, 5, 181–90.

[7] Milton K. Munitz, *Space, Time and Creation*, Glencoe, Illinois, 1957.

[8] R. Harré, 'Philosophical Aspects of Cosmology', *British Journal for the Philosophy of Science*, 1962, 13, 104–19.

I have not produced detailed counter-arguments to deal with each of these criticisms individually, but consider that I have met all the criticisms in this chapter and in Chapters 1 and 14 of this book. For a general attempt to show that many such criticisms are invalid see:

[9] G. H. Bird, 'The Beginning of the Universe', *Proceedings of the Aristotelian Society*, Supplementary Volume, 1966, 40, 139–50.

Index